I0820195

VII
SCHIRRA · EISELE · CUNNINGHAM

THE
APOLLO
PHOTO
ARCHIVE

APOLLO 7
IN PHOTOGRAPHS

VII
SCHIRRA · EISELE · CUNNINGHAM

THE
APOLLO
PHOTO
ARCHIVE

APOLLO 7 IN PHOTOGRAPHS

J.L. PICKERING AND JOHN BISNEY

WITH ED HENGEVELD

Other Schiffer books by the authors
Apollo 1 in Photographs: The Apollo Photo Archive
978-0-7643-7009-0

Library of Congress Control Number: 2025930652

Designed by Beth Oberholtzer
Cover design by Jack Chappell
Type set in Eurostile/Collier/Arno Pro

ISBN: 978-0-7643-7010-6
ePub: 978-1-5073-0584-3

Printed in India
10 9 8 7 6 5 4 3 2 1

Published by Schiffer Publishing, Ltd.
4880 Lower Valley Road
Atglen, PA 19310
Phone: (610) 593-1777; Fax: (610) 593-2002
Email: info@schifferbooks.com
Web: www.schifferbooks.com

Contents

Foreword

As the daughter of astronaut Donn Eisele, the most common question I get is "What was it like growing up with an astronaut for a father?" I have yet to come up with a good answer. To me, he was just Dad, like any other middle-class American father. He went to work, came home for dinner, and attempted to help with homework. I say attempted because he was so brilliant he could do any complicated math problem immediately in his head but had a difficult time explaining how he got the answer. He also loved airplanes. When I was eleven, he took me to an air show and talked a guy into letting him fly a biplane, and up we went to do loops in the air. The fact that his office happened to be at NASA, and one day his commute took him to space, seemed almost incidental in our day-to-day life.

Before they became known for their groundbreaking mission aboard Apollo 7, my father and his crewmates, Wally Schirra and Walt Cunningham, were three very different men brought together by a shared vision. Wally was the seasoned veteran; a Mercury and Gemini astronaut whose steady hand and quick wit made him the group's natural leader. Walt was a skilled pilot and engineer, with a deep understanding of spacecraft systems. And my father, Donn, brought his own brand of grit and technical expertise to the team. Their personalities were distinct, but together they complemented each other in a way that made them a formidable crew.

What truly captured my father's essence, both as an astronaut and as a person, was what the public knew as "The Wally, Walt, and Donn Show"—the first live televised broadcast from space, which earned them an Emmy. Those transmissions revealed not just the technical achievements of the mission but also the human side of space exploration, showing three professionals who could maintain their sense of humor and camaraderie even under the most-challenging circumstances.

This collection of photographs offers a unique glimpse into that historic mission. I'm reminded that while he may have been "just Dad" to me, he was part of something extraordinary. These images capture a pivotal moment in human history—the first successful crewed Apollo mission. They tell the story of three brave men who carried the weight of the space program on their shoulders after the tragic Apollo 1 fire and triumphantly demonstrated that the redesigned Apollo spacecraft was ready for its next critical steps. These aren't just documentation of a space mission; they're memories of a time when my father, along with his crewmates, helped write a crucial chapter in humanity's greatest adventure.

Kristy Eisele
Brooklyn, New York

Introduction

The first manned Apollo flight represented a fresh start to the program in many ways, and its success was the result of the efforts of thousands of NASA and contractor engineers, technicians, and support personnel to recover from the Apollo 1 tragedy the year before. By the fall of 1968, changes had been made to almost all aspects of the spacecraft after the fatal fire that claimed the lives of three astronauts during a countdown test at Launch Complex 34 (LC-34) on January 27, 1967, at Cape Kennedy Air Force Station (AFS), Florida.

To reduce fire hazards before and after liftoff, the command module (CM) cabin atmosphere was changed from 100 percent oxygen to 60 percent oxygen and 40 percent nitrogen (shortly after launch, however, the atmosphere would be slowly enriched, eventually becoming pure oxygen). The original two-piece hatch was replaced by a single hatch, which could quickly be opened from the inside. Materials in the CM, including the space suit outer layer, were replaced with less flammable substitutes such as Beta cloth, a fabric made of Teflon-covered woven silica fiber developed by DuPont that would not burn.

Three unmanned Apollo test missions had flown since the Apollo 1 accident, including the first two flights of the Saturn V booster, intended to send astronauts to the Moon. As an Earth-orbiting mission with no lunar module (LM), however, Apollo 7 would use a smaller Saturn IB. On August 19, 1968, the National Aeronautics and Space Administration (NASA) announced that October 11 would be the launch date for Apollo 7, intended, as Apollo 1 had been, to be a shakedown flight for the command and service modules (CSM).

The mission launched from LC-34 at 11:03 a.m. EDT on October 11. With all subsequent manned flights using LC-39 at NASA's adjacent Kennedy Space Center (KSC), it was the last manned launch to use a pad at the US Air Force (USAF) installation.

Walter M. "Wally" Schirra Jr., forty-five, was commander, becoming the only astronaut to fly on Mercury, Gemini, and Apollo missions. He was joined by rookies Donn F. Eisele, thirty-eight, a USAF major, as CM pilot, and civilian R. Walter Cunningham, thirty-six, as LM pilot. The new crew designations were based on lunar missions; in reality, Eisele became expert on the navigation system, and Cunningham, with no LM, focused on CM systems.

On the second day, Apollo 7 rendezvoused with the Saturn booster's second stage. The astronauts approached to within about 70 feet to simulate a rescue by the Apollo command ship of a stranded LM in orbit around the Moon. This was also a key test of the SM's Service Propulsion System (SPS), which would be needed to enter and leave lunar orbit.

The flight included two firsts for American spaceflight: the first live onboard television, and the first time there was noticeable friction between a commander and Mission Control in Houston. With a long mission largely devoted to systems checks, flight planners had also added experiments to their workload, scheduling almost every free minute. The tight timeline, combined with Schirra developing a head cold, led to onboard frustration and short tempers. Schirra once complained about flight controllers not allowing him enough sleep. He took aspirin and decongestant tablets but had symptoms for much of the mission. Hot meals and a much-roomier spacecraft, however, made the flight more comfortable than Mercury and Gemini flights.

Although Gordon Cooper had experimented with a "slow-scan" TV camera during his 1963 Mercury flight, the picture quality was poor, and the images were shown on a delayed basis. For Apollo, the RCA Corp. had developed a 4.5-pound black-and-white camera that, while still a slow-scan unit, provided much-better images (which had to be converted on the ground for use by broadcasters).

Live television was transmitted to ground stations on seven occasions, enlivened by the crew with hand-lettered signs including "From the lovely Apollo Room high atop everything" and "Keep those cards and letters coming in, folks." The latter prompted entertainer Dean Martin to send a telegram to Mission Control, reminding the crew that he used that slogan on his weekly TV show. Martin added a postscript: "Like all Americans, I am proud of you," and a PSS, "I was higher last night than you are now." Though Schirra originally criticized the camera as being an "inferior product," with an embarrassing picture quality, the telecasts became a mission highlight.

In the last few days of the flight, the congested crewmen became concerned about wearing their helmets during reentry, which would prevent them from blowing their noses to relieve pressure on their eardrums during the change in altitude. Mission Control tried to persuade them to wear the helmets anyway, but Schirra refused. They took decongestant pills and landed in the North Atlantic without any ill effects on October 22, 1968. The broadcast networks provided the first live color TV coverage of a recovery from USS *Essex*, the prime recovery ship, but overcast skies limited visibility.

At nearly eleven days, Apollo set the stage for a much more ambitious plan. According to NASA's original schedule, Apollo 7 would have been followed by testing the Apollo spacecraft with an LM in Earth orbit. Since LM development was behind schedule, however, officials began considering a bold move: if Apollo 7 experienced no significant problems, they would send Apollo 8, but without an LM, into lunar orbit. Apollo 7's success prompted NASA to publicly give Apollo 8's Moon flight the green light just three weeks after 7's splashdown.

Acknowledgments

Matthew Beddingfield
Shane Bell
Howard Benedict
Mark Gray
Ken Havekotte
Juliana Shepard Jenkins
William Killen
Alan Lawrie
Robert Pearlman
Scott Schneeweis
Ron Specht
Myra Taub Specht
John Uri
Mark Usiack
Thomas Usiack
Beth Voecks
Ron Woods

CHAPTER 1

The Astronauts

US Navy captain Walter M. "Wally" Schirra holds docked Gemini and Agena models in an unreleased 1964 portrait.

The Apollo 7 mission emblem was designed by Apollo spacecraft prime contractor North American Rockwell (NAR) graphic artist Allen Stevens.

US Air Force (USAF) major Donn F. Eisele's 1964 unreleased portrait

R. Walter "Walt" Cunningham's 1964 unreleased portrait. A civilian astronaut, he served in the US Marine Corps Reserve as a colonel.

Schirra's first NASA portrait in 1959

Schirra poses with the other six original astronauts and flight director Chris Kraft in the Mercury Control Center at Cape Canaveral Missile Test Annex, Florida, on September 22, 1960. *Left to right*: Schirra, Deke Slayton, Gus Grissom, Kraft, Gordon Cooper, Scott Carpenter, John Glenn, and Alan Shepard.

Schirra and the other Mercury astronauts are welcomed to Houston, Texas, with a parade on July 4, 1962; the placard mistakenly has his middle initial as "A." He arrived just hours earlier from the Cape after briefings on his upcoming flight. Schirra's family did not join him because his wife was recovering from a broken leg.

Schirra is assisted on launch morning by NASA suit technician Alan Rochford (*left*) on October 3, 1962, in Hangar S at Cape Canaveral Air Force Station (AFS), Florida.

Schirra poses with a Mercury model before his six-orbit Mercury-Atlas (MA)-8 mission. *Bill Taub archive*

Atlas 113D launches Sigma 7 with Schirra aboard into orbit at 8:13 a.m. EST on October 2, 1962, from LC-14 at Cape Canaveral AFS.

Schirra takes a piece of meat from USAF instructor H. Morgan Smith as other members of the first two groups of astronauts sample the fare during jungle survival training at the Caribbean Air Command's Tropic Survival School at Albrook AFB in the Panama Canal Zone, on June 3, 1963.

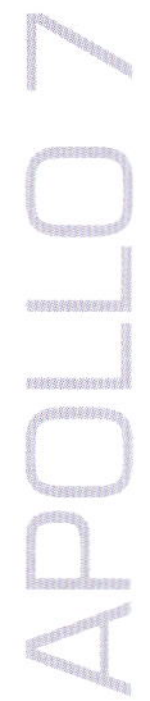

Left to right: Schirra, Al Chidester of the US Geological Survey, Frank Borman, and Ed White take part in geology training at the Grand Canyon in Arizona from March 6 to 7, 1964. Schirra, a camera buff, has a Hasselblad 500C camera like the one he used on Sigma 7.

Gemini VI astronauts Schirra (*right*) and Tom Stafford pose for a crew portrait at the Kennedy Space Center (KSC) on October 20, 1965. They would spend just more than one day in orbit when their mission was changed to rendezvous with Gemini VII because their Agena target vehicle had exploded after launch.

Gemini VI heads into orbit atop its Titan II booster at 10:00 a.m. EST on December 15, 1965, from Launch Complex (LC)-19 at Cape Kennedy AFS).

Gemini VII is photographed in orbit from Gemini VI on December 15, 1965.

Eisele's first NASA portrait in 1963

Eisele, Chidester, and astronaut Ted Freeman chat during geology training at the Grand Canyon from March 6 to 7, 1964, during an outing described as "mainly show and tell with questions" covering basic geologic structures and rock types.

Eisele and Cunningham, among the third group of astronauts selected, tour the LC-37 blockhouse during their first trip to Cape Kennedy AFS on April 9, 1964. *Left to right*: unidentified, Edwin Aldrin, Alan Bean, Roger Chaffee, Bill Anders, Russell "Rusty" Schweickart (*behind Anders*), Eisele, Gene Cernan, Ted Freeman, Cunningham, Group 1's Schirra (*behind tour guide*), C. C. Williams, and Mike Collins.

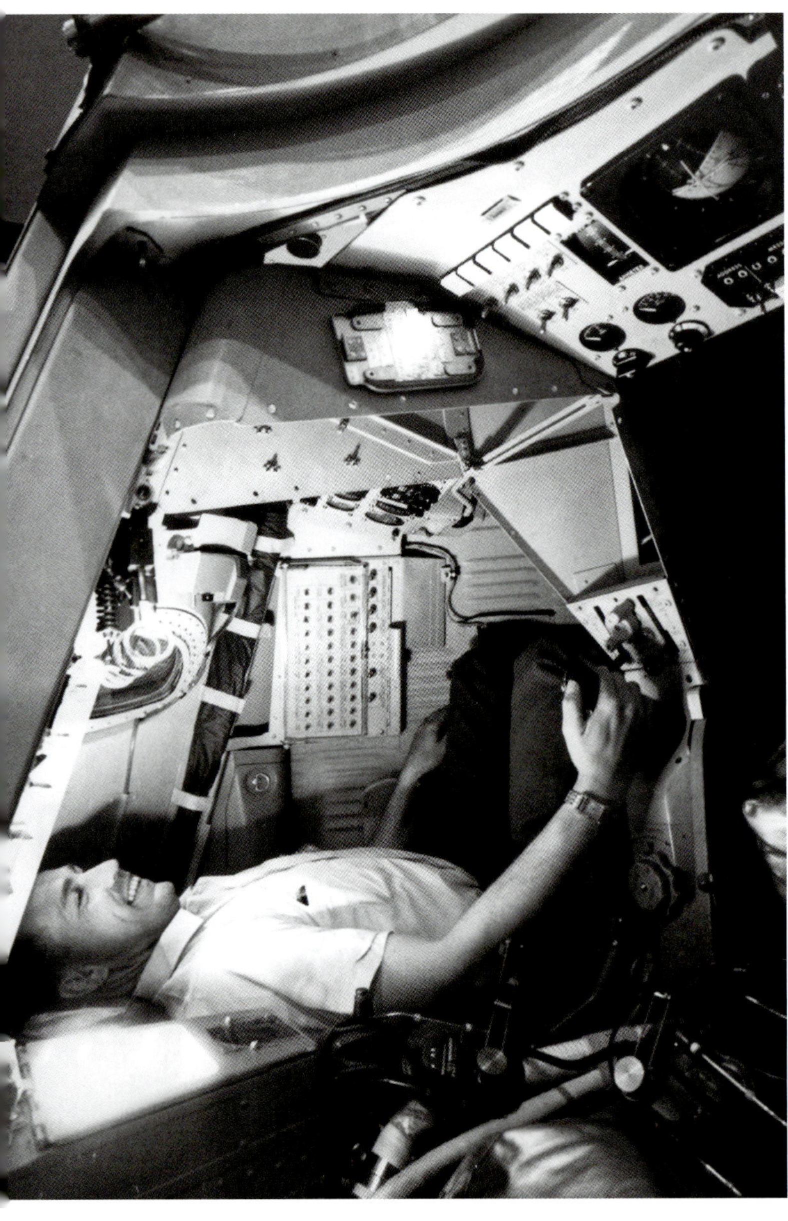

Eisele tries out the Gemini Mission Simulator in the Mission Control Center during the Cape visit.

Eisele (*center*) watches Choco Indian chief Zarco prepare food during jungle survival training near Albrook AFB in Panama from June 22 to 26, 1964. *Bill Taub archive*

Eisele inspects reptile specimens on display during the jungle training.

Eisele and six fellow astronauts wait to board a plane to Hawaii for geology training on January 11, 1965. *Left to right*: Eisele, Dave Scott, C. C. Williams, Chaffee, and Bean. In front are Dick Gordon and Pete Conrad. *Bill Taub archive*

Eisele examines a rock sample on Hawaii's "Big Island" during geology training from January 11 to 15, 1965. The trip is to observe basaltic volcanic features. Fresh, recent, and ancient lava-flow surfaces could be compared to possible lunar surface features.

Eisele enters instructions into an Apollo command module (CM) Display Keyboard next to the Primary Guidance, Navigation and Control System demonstrator during a visit to the MIT Instrumentation Laboratory in Cambridge, Massachusetts, in March 1965. *MIT photo*

Cooper and Eisele watch the launch of AS-201 on February 26, 1966, from LC-34 at Cape Kennedy AFS. It is the first test flight of an Apollo command and service module (CSM) using a Saturn IB booster. The suborbital flight is a partially successful demonstration of the service propulsion system and the reaction control (RCS) systems of both modules and shows the CM's heat shield could withstand reentry from Earth orbit.

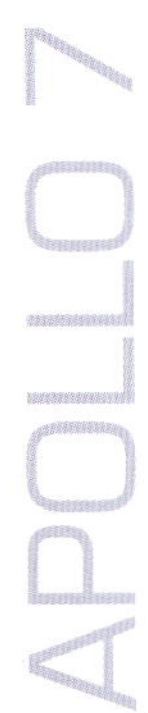

Cunningham's first NASA portrait in 1963

Cunningham, Gordon, and Schweickart chat near the LC-34 service structure during a visit to Cape Kennedy AFS on April 9, 1964.

Group 3 astronauts pose at the top of LC-37 on April 9. *Left to right*: Aldrin, Schweickart, Freeman, Anders, Charles Bassett, Williams, Bean, Cunningham, Collins, Scott, Chaffee, Cernan, and Gordon.

MSC geologist Uel Clanton (*left*) with Williams, Cernan, Anders, Schweickart (*behind Anders*), Cunningham, and Freeman during geology training in Flagstaff, Arizona, from April 20 to May 2, 1964. They also visit Kitt Peak Observatory.

Left to right: Cunningham, Al Chidester of the US Geological Survey, and Neil Armstrong at the Philmont Boy Scout Ranch in Cimarron, New Mexico, on June 4, 1964. This trip involves more-advanced geology training than previous sites.

Cunningham holds a flare during jungle survival training at Albrook AFB in Panama, from June 22 to 26, 1964. *Bill Taub archive*

Cunningham (*left*) and Bean take a meal during the training. *Bill Taub archive*

Cunningham wears a prototype space suit during tests at the Yapoah lava flow at McKenzie Pass in Bend, Oregon, on August 25, 1964. His faceplate fogged up, however, and he stumbled and fell, tearing a hole in the suit on the jagged lava, which was patched up with duct tape.

Cunningham practices using a hand mirror to signal rescue aircraft during desert survival training at Carson Sink north of Stead AFB, Nevada, on August 12, 1965.

Anders places a horned toad on Cunningham's shoulder near Stead AFB. Stead was home to the USAF's Survival, Evasion, Resistance, and Escape School.

CHAPTER 2

September 1966–December 1967

The crew for Apollo 2, with Schirra (*front*) as command pilot, Eisele (*right*) as senior pilot, and Cunningham as pilot, is named on September 29, 1966. Their flight is intended to be a repeat of Apollo 1, but with more emphasis on science. The crewmen pose the day before at MSC in front of the first Apollo Mission Simulator (AMS), built by the Link Group of General Precision Systems of Sunnyvale, California.

Left to right: Schirra, Eisele, and Cunningham in the hatch of a CM mockup at MSC on September 28

Cunningham (*left*) and Schirra wait to enter CM-014 in Building 290 at prime contractor North American Aviation (NAA) in Downey, California, for tests on October 25, 1966. The forward access tunnel is in the foreground. Intended to be Apollo 2, this Block I CM would instead be sent to KSC the following February to aid in the Apollo I fire investigation.

NAA developed two versions of the CM under a "building block" approach. Block I spacecraft were to be test flights in Earth orbit, while the Block II versions would be for lunar missions. Two unmanned Block I CMs were launched with Saturn I boosters in 1966, and two manned Block I flights—Apollo 1 and Apollo 2—were planned but canceled after the Apollo 1 fire. The Block II spacecraft then underwent significant safety modifications—its hatch was redesigned, its cabin atmosphere at launch was changed to 60 percent oxygen and 40 percent nitrogen, and flammable material was replaced or removed. The first Block II manned mission would be Apollo 7. Beginning with Apollo 9, it would also carry a docking mechanism for the lunar module (LM).

Schirra prepares to enter CM-014 with Cunningham behind him. The astronauts wear a modified Gemini GC3 suit, the A1C. This version added new electrical and environmental connectors, and a helmet shell protected the Plexiglas visor when open. It was the final suit made by the David Clark Co. of Worcester, Massachusetts.

Eisele and Cunningham get settled inside the CM, with Schirra out of frame at left. Their planned Apollo 2 mission, however, is canceled three weeks later, and they are reassigned as the Apollo 1 backup crew. Schirra later says he was disappointed but concurs in the decision.

NAA technicians surround CSM-101 in workstand 2A at Downey on November 29, 1966; the CM's aft heat shield has not yet been attached. This first flight-qualified Block II CM would eventually fly as Apollo 7.

Eisele inflates water wings as the first crewman to climb out of Boilerplate 1101A (BP-1101A) in the Water Immersion Facility in Building 5 at MSC during recovery training for Apollo 1 on December 5, 1966.

Schirra prepares to exit the CM boilerplate to join Eisele (*left*) and Cunningham in the life raft.

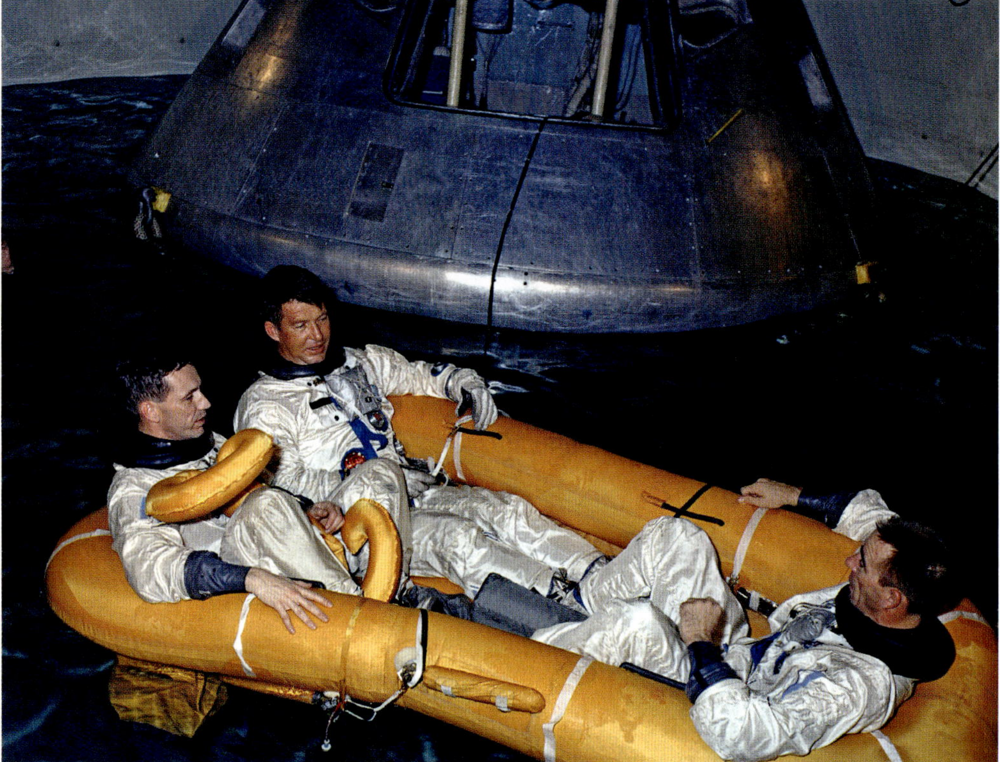

Eisele, Schirra, and Cunningham occupy the raft. They would repeat this training the next day in the Gulf of Mexico off Galveston, Texas.

The Apollo 7 astronauts meet with the news media at NAA on May 10, 1967. NASA administrator Jim Webb had announced their selection the day before during Senate testimony. *Left to right*: backup crewmen Gene Cernan, John Young, and Tom Stafford, and prime crewmen Cunningham, Eisele, and Schirra. "It is time for self-recrimination [about Apollo 1] to end," said Schirra. "We have a job to do."

Left to right: Cunningham, Eisele, and Schirra pose with a Block II CM mockup, which showcases the new outward-opening one-piece hatch. Some NAA workers were still worried, however, that their company might lose the Apollo contract.

The crewmen share a laugh. *Front, left to right*: Cunningham, Eisele, and Schirra; *back, left to right*: Cernan, Young, and Stafford. The astronauts are at NAA to go over the safety improvements to the spacecraft.

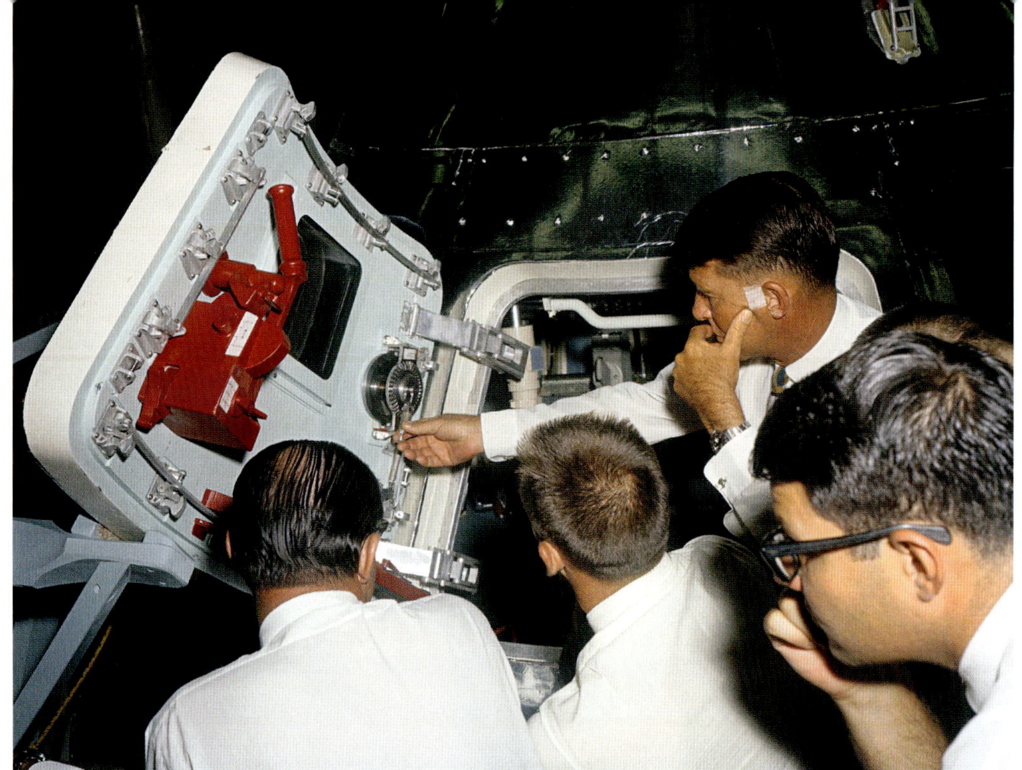

Schirra inspects a model of the new hinged unified hatch in a CM mockup in Building 290 during a visit to NAA on June 29, 1967. His hand is on the vent valve, which could also be operated from the outside and could vent the cabin pressure in one minute. The hatch was opened by the single large actuator handle (here painted red) with a ratchet mechanism, which operated fifteen latches. A crewman would pump the handle five times for normal operation, but in an emergency, a pressurized gas-driven counterbalance would open the hatch in seconds. Cunningham watches below Schirra.

Schirra grips the rotational hand controller as he peers out of the rendezvous window from the commander's couch in the Command Module Procedures Simulator (CMPS) in Building 5 at MSC in August 1967. He looks through a combining glass normally between the Crewman Optical Alignment Sight and the window, used to line up the CM for rendezvous with the LM.

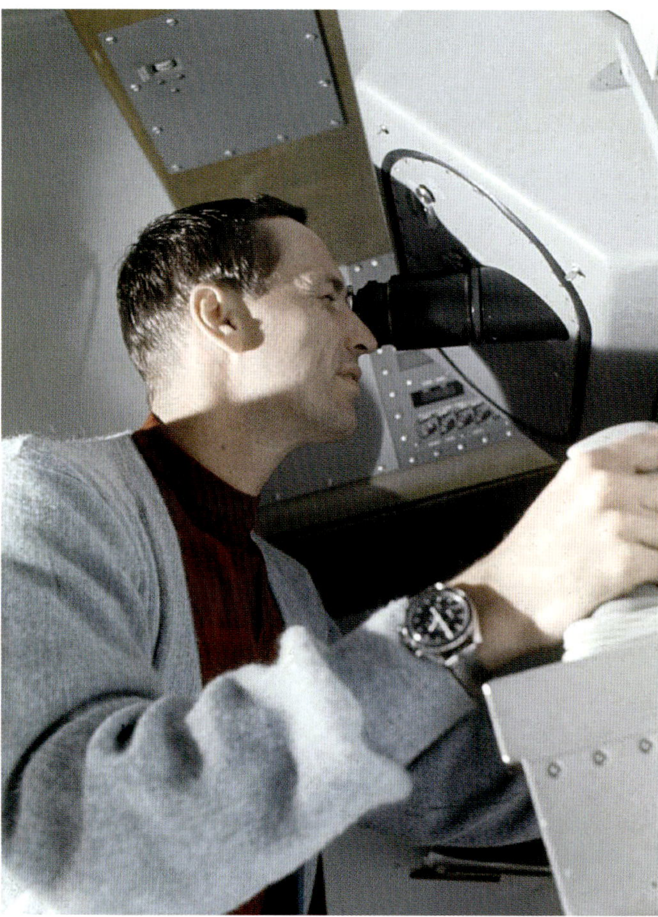

Eisele sights through the eyepiece of the Guidance, Navigation and Control System in the CMPS's lower equipment bay. He would joke, "I'm the navigator on this flight, and I have a right to know where we are."

The Apollo 7 crew poses with an S-IVB stage at the McDonnell Douglas Space Systems Center in Huntington Beach, California, on September 5, 1967. *Left to right*: Eisele, Schirra, Cunningham, and N. B. Robins, assistant to the Saturn/Apollo program's senior director. Douglas, which built the second stage of Apollo 7's Saturn IB booster, had merged with McDonnell Aircraft in April.

Apollo 7's S-IVB-205 (*left*) is parked outside the McDonnell Douglas Vehicle Checkout Laboratory in Rancho Cordova, California, on September 28, part of the company's Sacramento Test Operations site. The other two stages are 206 and 207, which would be used for Saturn IB Skylab launches in 1973.

With the Apollo program's final sequence taking shape, program director Maj. Gen. Sam Phillips, USAF, decided it was time to simplify the public nomenclature for each mission, although NASA still used its system internally. On April 24, 1967, the associate administrator for space flight, George Mueller, approved a plan to retroactively designate AS-204, the fatal pad fire, as Apollo 1. The first Saturn V launch, AS-501, would be renamed Apollo 4 (Mueller skipped Apollo 2 and 3). Apollo 5 would be the only unmanned LM test flight using a Saturn IB, and Apollo 6 would be the second unmanned Saturn V test launch.

The first Saturn V heads skyward from LC-39A at KSC at 7:00 a.m. EST on November 9, 1967. The unmanned flight, formally known as AS-501 but dubbed Apollo 4, is a shakedown mission for the huge booster, the most powerful rocket ever launched.

Apollo 4 heads out over the Atlantic Ocean during the first few minutes of flight.

Cunningham (*left*) and Schirra monitor the Apollo 4 flight in the Mission Operations Control Room (MOCR) at MSC's Building 30 in Houston.

One of the 755 Earth views captured by a 70 mm automatic camera aboard the CM, including the highest-altitude photos of the planet ever taken.

Schirra (*left*) chats with Dr. Charles Berry in the MOCR during Apollo 4, who serves as a flight surgeon for several manned missions.

The Apollo 4 CM is brought aboard the prime recovery ship, the aircraft carrier USS *Bennington*, after three orbits. It splashes down nine hours after launch about 10 miles from the target northwest of Midway Island in the Pacific Ocean, and the mission is judged a major success.

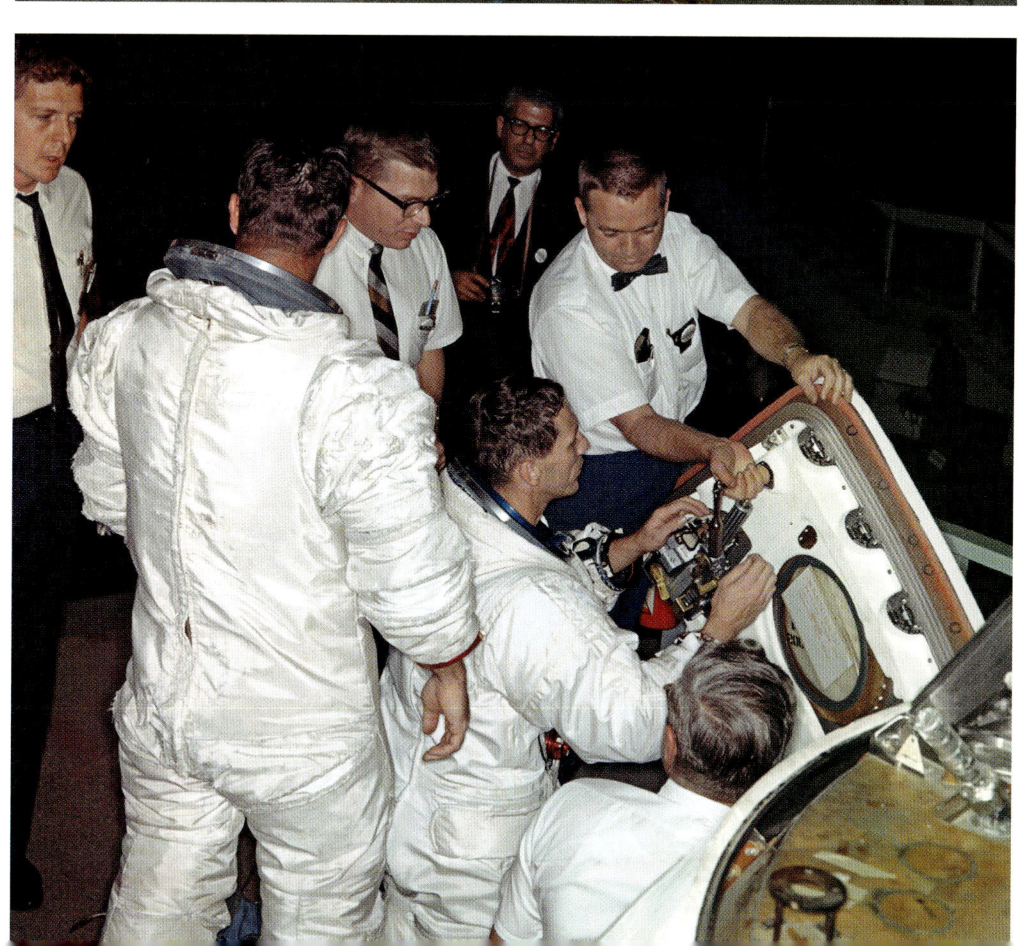

Eisele (*center*) and Schirra (*left*) inspect the new quick-opening CM hatch at North American Rockwell (NAR) on November 16, 1967, as managers look on (NAA had merged with Rockwell-Standard that September to form NAR).

Eisele (*foreground*) is the only suited crewman with Schirra (*center*) and Cunningham (*top*) inside a CM at NAR on November 16.

Eisele visits with workers at AC Electronics in Oak Creek, Wisconsin, in December. The company assembles and tests the CM's guidance and navigation system and its inertial measurement unit.

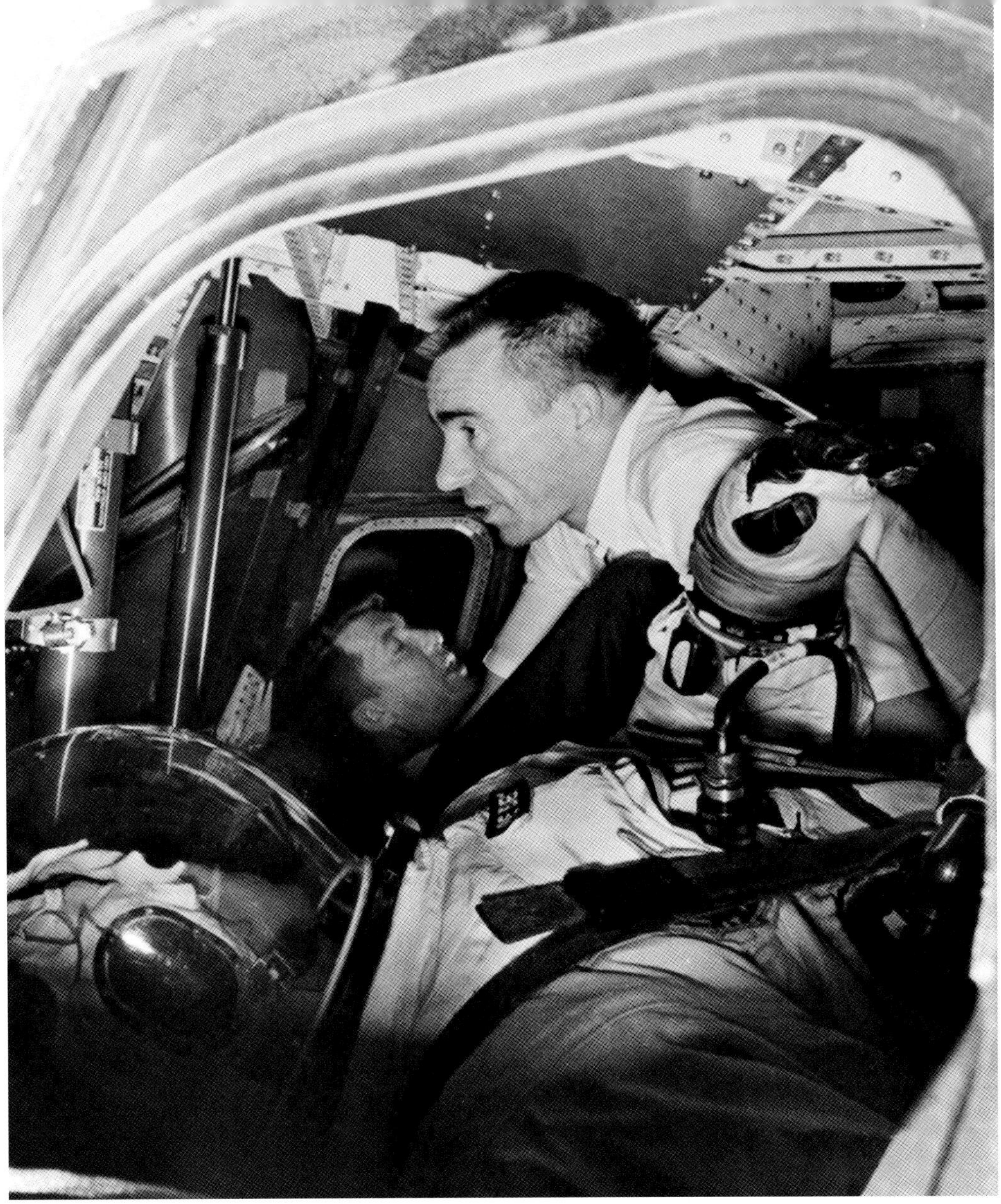

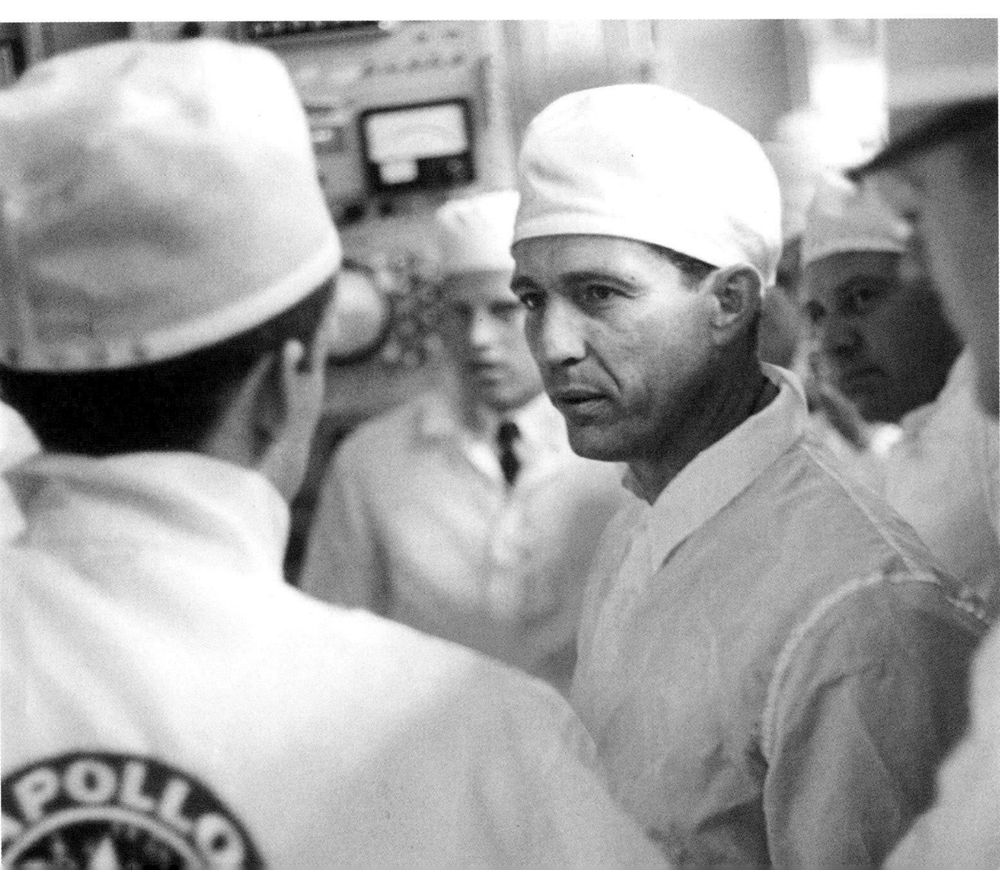

On December 8, the astronauts visit NAR in Downey to review equipment and inspect their CM.

Schirra (*center*) examines a Hasselblad camera. He is flanked by support crew member Ron Evans to his right and Eisele to his left. Backup crewman Gene Cernan is behind Eisele at right.

Director of flight crew operations Deke Slayton (*left*) reviews CM-101 stowage with an NAR manager.

Schirra and Cunningham (*bottom*) are briefed on the new unified hatch's operation by an NAR manager. NASA's Apollo CSM manager Ken Kleinknecht, *standing at center*, would chair a meeting with the contractor four days later to go over outstanding issues.

Schirra, Cunningham, and Eisele wear A6L space suits and communication carriers made by the International Latex Corp. (ILC) of Dover, Delaware, before a test aboard CM-101. Behind them is suit tech Keith Walton.

Schirra prepares to board and will occupy the left-hand couch as commander.

Eisele, who will occupy the center couch, boards last.

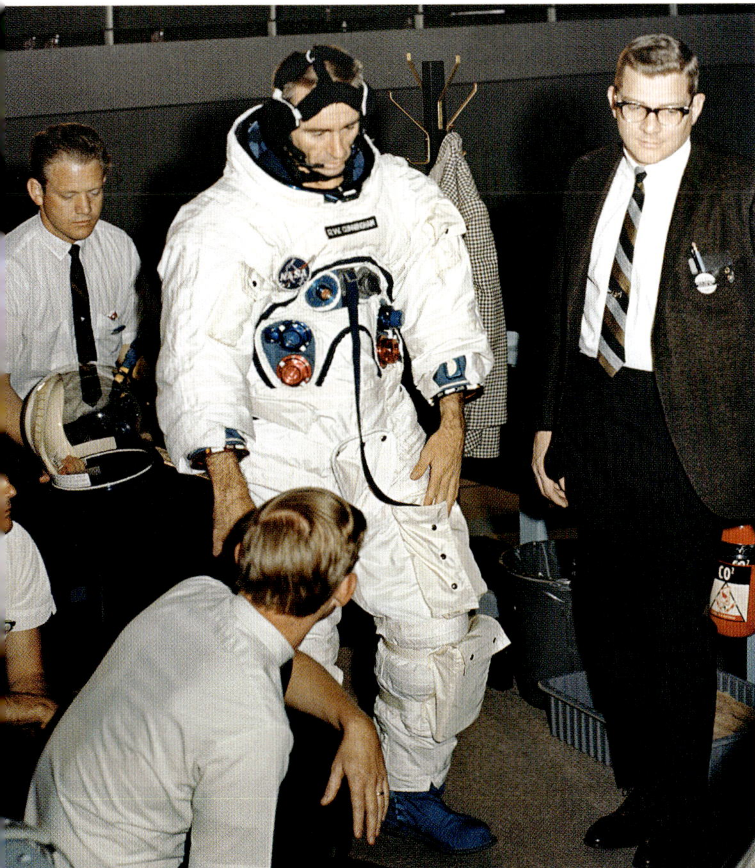

Cunningham prepares to ingress and will ease into the right-hand couch.

The astronauts are strapped in.

Schirra begins to egress after the four-hour session.

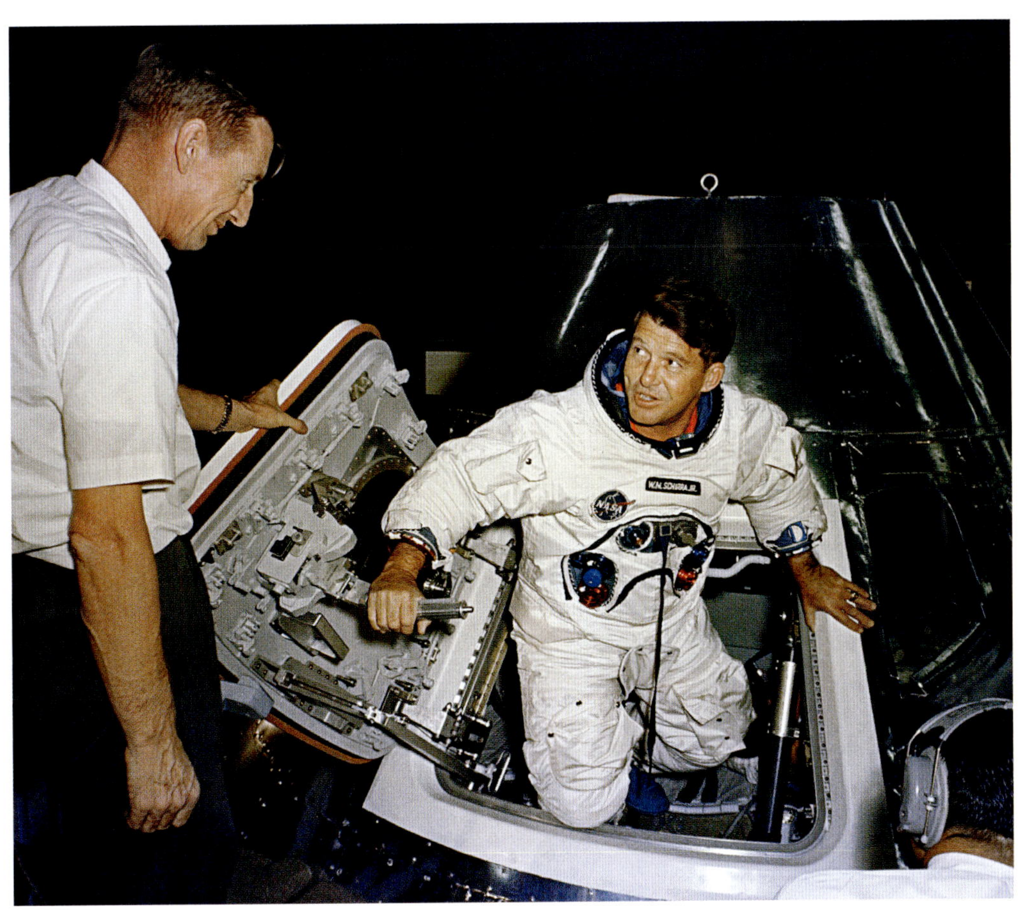

NASA suit technician Clyde Teague stands by to help Schirra.

CHAPTER 3

January–April 1968

Schirra (*left*) and Cunningham chat near the LC-37 blockhouse in January as the Saturn IB behind them is prepared to launch Apollo 5 (AS-204), the only unmanned test flight of the lunar module (LM) in Earth orbit.

Workers secure the LM for Apollo 5 (LM-1), festooned with red "Remove Before Flight" streamers, in its Spacecraft / Launch Vehicle Adapter (SLA) on November 11, 1967, in the Manned Spacecraft Operations Building (MSOB) in the KSC Industrial Area.

Apollo 5 thunders off the pad at 5:48 p.m. EST on January 22, using the Saturn IB booster originally erected on LC-34 in August 1966 to launch Apollo 1 in February 1967. The successful mission demonstrates the LM's ascent and descent propulsion systems and the ability to abort a Moon landing and return to lunar orbit.

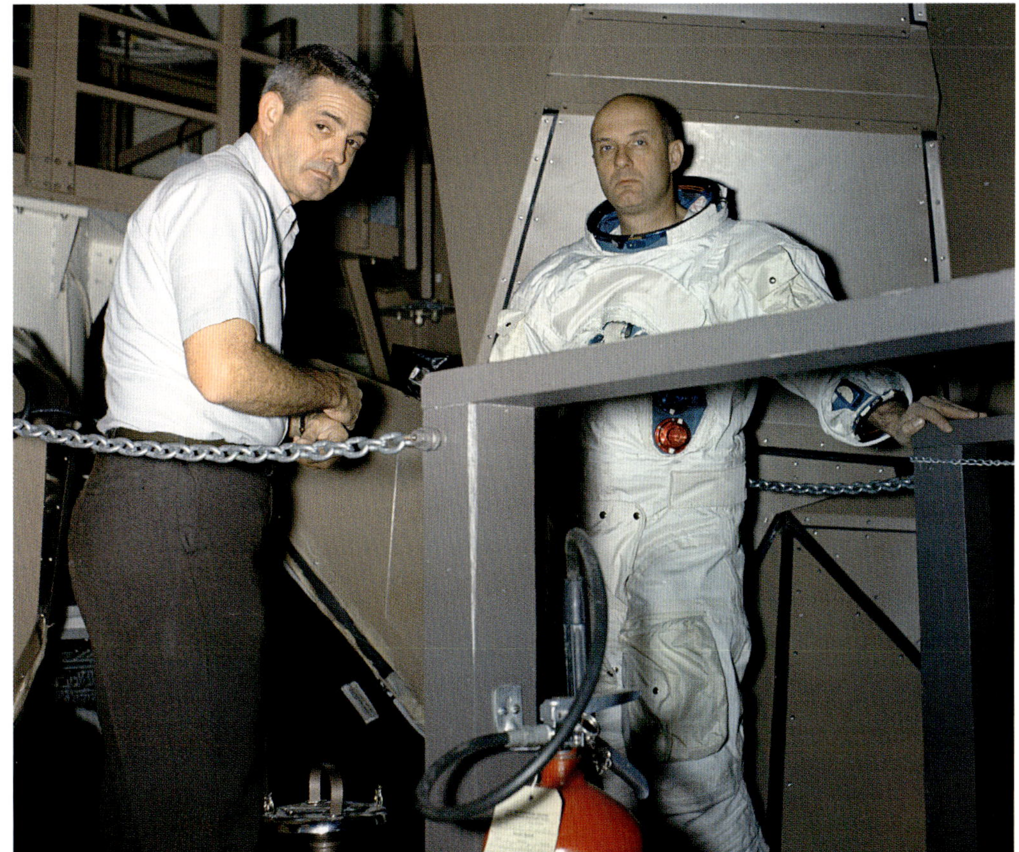

Apollo 7 backup commander Tom Stafford (*right*) is ready for a session in the Apollo Mission Simulator (AMS) behind him in Building 5 at MSC on January 23. At left is suit technician Walter Salyer of MSC's Crew Systems Division.

Apollo 7 backup LM pilot Gene Cernan holds his helmet before the simulator training.

Apollo 7 backup CM pilot John Young wears an early communications carrier (headset) before boarding the AMS; Stafford and Cernan are already inside with Salyer.

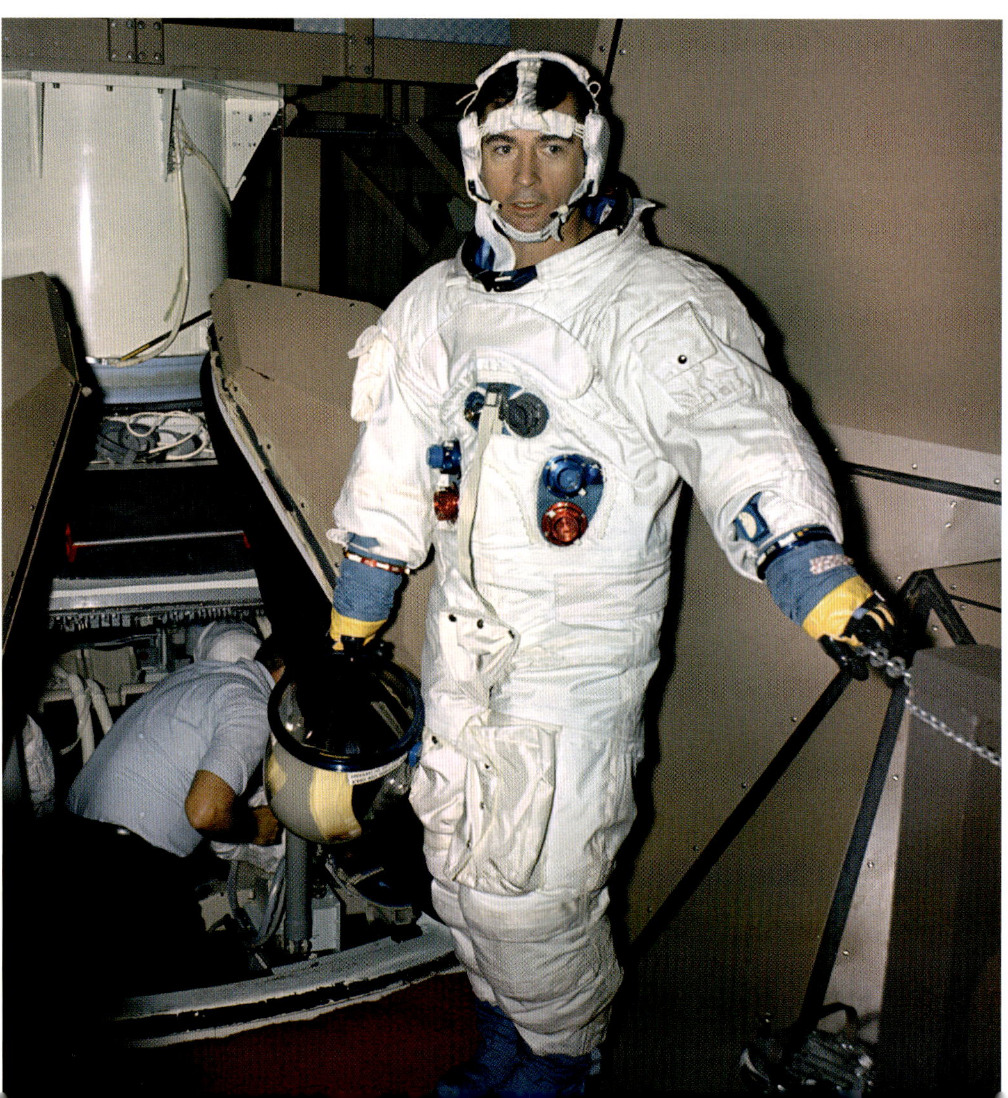

Left to right: Schirra, Cunningham, and Eisele pose with an Apollo/Saturn IB model outside MSC on February 13. Schirra holds the Hasselblad still camera, and Eisele holds a Maurer 16 mm data acquisition camera that will be used on the mission.

Left to right: Eisele, Cunningham, and Schirra with a CSM model

Cunningham (*left*) and Eisele look over the modified 70 mm Hasselblad 500C camera, which uses an 80 mm lens, during the photo shoot. It will be used primarily for weather and terrain photography.

Eisele sights through the Hasselblad frame viewfinder. MSC's Richard Underwood (*right*) trains the astronauts in using photographic equipment.

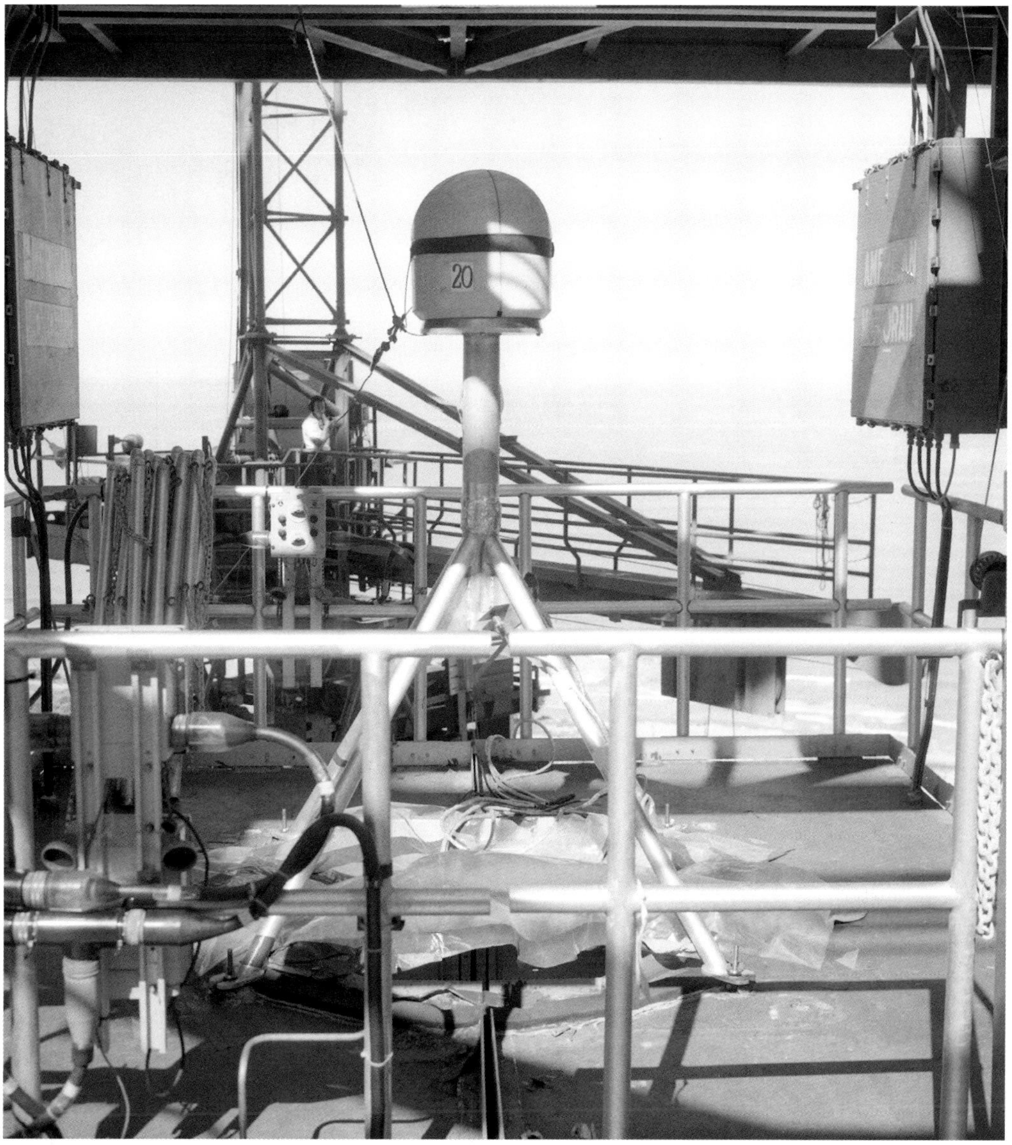

A technician standing on the LC-34 umbilical tower crane in the background tests the system for uncovering the Q-ball on March 20. The lower cable is attached to a mechanism to cut the 2-inch-wide rubber band holding the two halves of the Styrofoam cover together seconds before launch. The Q-ball would be mounted at the top of the CM's Launch Escape System (LES) to measure differences in aerodynamic forces, symbolized by Q in flight equations, during flight.

On March 28, a NASA tugboat pulls a KSC barge carrying the S-IB first stage for Apollo 7 through the Canaveral Barge Canal, heading west from Port Canaveral toward the Banana River. A USAF pusher ship follows. The stage had been transferred at the port from the larger barge *Point Barrow* after a voyage from the Michoud Assembly Facility in New Orleans, Louisiana, through the Gulf of Mexico and around Florida.

The barge passes under one of the twin double-leaf bascule drawbridges before the Canaveral Lock.

The tug, barge, and pusher approach the Canaveral Lock at the west end of the barge canal. The largest navigation lock in Florida was built in 1965 by the US Army Corps of Engineers to allow passage of large vessels between Port Canaveral and the Banana River.

The vessels approach the Cape Kennedy AFS Industrial Area wharf on the Banana River, about 7 miles north of Port Canaveral.

The barge swings in toward the dock.

The S-IB, showing the outline of its eight H-1 engines under the neoprene-coated nylon tarp, will be carefully backed off the barge. The KSC Transportation Branch, staffed by contractor Bendix, coordinating the arrival, is responsible for operating everything from custom equipment trailers, barges, and administrative aircraft to a taxi fleet and shuttle buses.

The NASA tractor will next be unhitched and moved to the aft end of the stage.

The tractor prepares to back the stage out and take it to Hangar AF, about 400 yards away, for inspection.

The second and final Saturn V test flight, Apollo 6 (AS-502), launches from LC-39A at 7:00 a.m. EST, on April 4. It carries an unmanned Block I spacecraft (CM-020 and SM-014) and a simulated LM (LTA-2R).

Excessive g-forces caused by unexpectedly severe "pogo" vibrations in the first stage caused pieces of the SLA to come off 133 seconds into the flight, captured by the Airborne Light Optical Tracking System. Two second-stage engine shutdowns then leave the CSM in a less-than-nominal elliptical orbit, and the S-IVB engine later fails to restart.

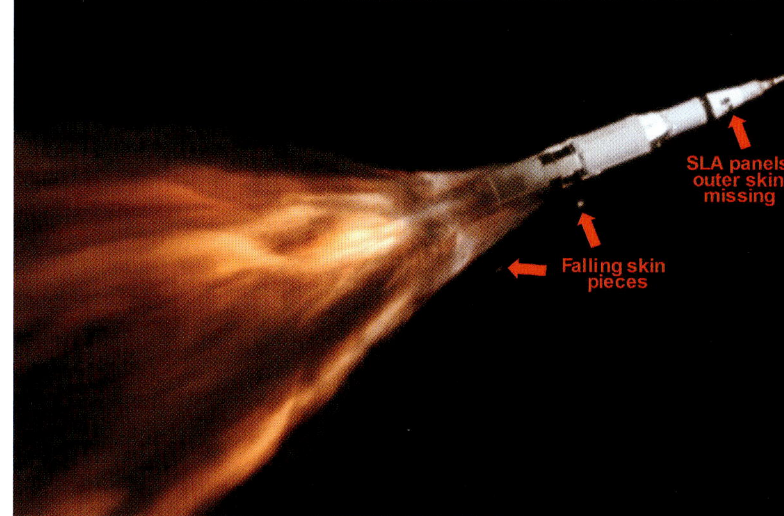

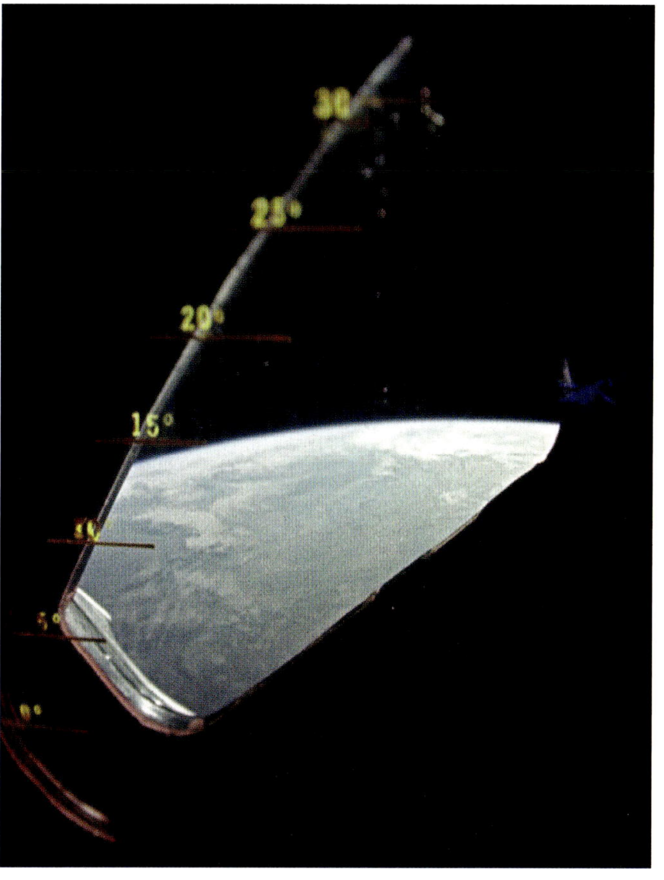

Earth's limb as photographed through a CM rendezvous window by a 16 mm camera aboard Apollo 6

Divers steady the Apollo 6 CM after splashdown north of Hawaii in the Pacific Ocean, almost ten hours after launch. Its uprighting bags are partially deflated. The CM is soon brought aboard USS *Okinawa*, and engineers at the Marshall Space Flight Center (MSFC) in Huntsville, Alabama, get to work to understand and resolve the booster's pogo and engine problems.

Astronaut Jim Lovell climbs into CM-007A aboard Motor Vessel (MV) *Retriever* in Galveston Bay, Texas, on April 4, for a forty-eight-hour exercise conducted by MSC's Landing and Recovery Division in case a CM could not be immediately recovered. *Retriever* is a World War II–era Landing Craft Utility transferred to NASA from the US Army, used for astronaut training during Gemini and Apollo from 1963 to 1972.

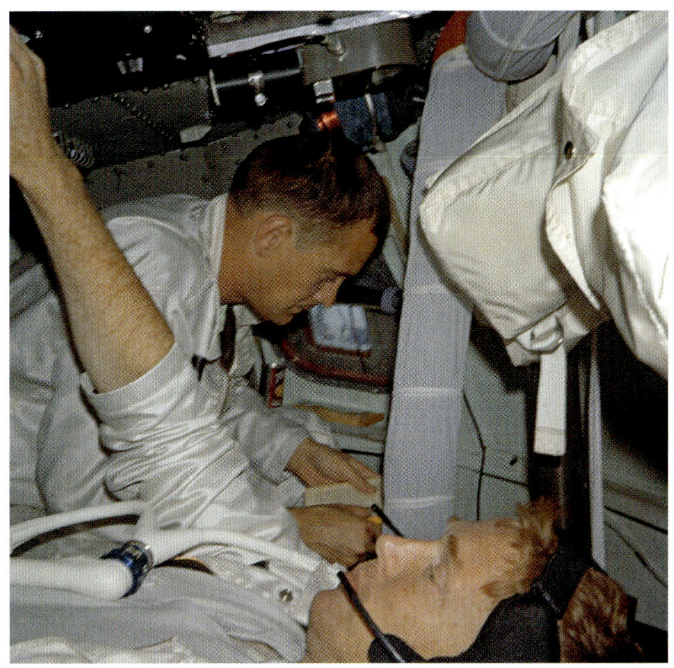

Astronauts Charlie Duke (*top*) and Stu Roosa joined Lovell aboard CM-007A. They encounter no serious habitability problems during the two days, although they did "not recommend the Apollo spacecraft for any extended sea voyages."

Retriever placed the CM apex down in Galveston Bay, and its uprighting bags inflated within six minutes. Pressure checks are made on the bags every four hours throughout the test.

McDonnell Douglas workers unload S-IVB-205, the second stage for Apollo 7, from the Super Guppy aircraft on April 7 at the Cape Kennedy AFS Skid Strip. It has arrived from the McDonnell Douglas Test Facility in Sacramento, California.

A McDonnell Douglas tractor pulls the S-IVB's trailer away from the aircraft. The four cannisters with hoses are part of the stage's environmental-control system during air transportation.

A tractor, hooked to the opposite end of the trailer, tows the S-IVB to the Vehicle Assembly Building (VAB) at KSC for inspection.

Just three days later, the Super Guppy delivers the Instrument Unit (S-IU-205) for Apollo 7's Saturn IB from MSFC. It houses systems for guidance and vehicle performance from liftoff until after orbital insertion.

The IU, manufactured by IBM in Huntsville, Alabama, is moved out from the Super Guppy atop its wheeled carrier.

Nearly 23 feet in diameter, the IU is the booster's nerve center, responsible for originating guidance commands for stage steering, engine ignition and cutoff, staging, primary timing signals, and telemetry. It will be taken to Hangar AF for inspection and acceptance.

The S-IB first stage, built by Chrysler Corp.'s Space Division at Michoud for Apollo 7, arrives at LC-34 on April 15.

Technicians prepare to remove the tarp from the stage. A flame deflector is in the background (*at left*), with LC-37 in the distance.

Service structure cranes begin to raise the first stage.

Workers steady the first stage with taglines as it goes vertical.

The pad crew pauses before the stage is moved into place above the launch pedestal; Chrysler workers on the forklift platform attach a tie line to a fin-mounting point.

A bridge crane moves the first stage toward the launch pedestal.

The first stage of the Apollo 7 Saturn IB is slowly lowered into place; two swing arms are at left.

The next day, a mobile crane begins to lift the aft end of the S-IVB from its transporter after the stage arrives at LC-34; the bridge crane is attached to the forward end.

Workers steady the stage with taglines as the bridge crane tilts it vertically, with its handling fixture still protecting its J-2 engine at left.

The S-IVB rests on its unfolded handling fixture.

The S-IVB is lowered toward the interstage.

The S-IVB has been mated to the interstage.

The combined second-stage assembly is raised into position.

On April 16, the S-IVB is moved horizontally toward the first stage of the Saturn IB.

Cranes lower the S-IVB over the first stage for mating.

Workers in the MSOB prepare to move a boilerplate service module (BP-30) and SLA on April 16 to LC-34 to serve as a stand-in until the flight CSM arrived. It saves time by checking out procedures and the interfaces between the spacecraft and ground equipment at the pad, since CSM-101 would not be ready for several weeks.

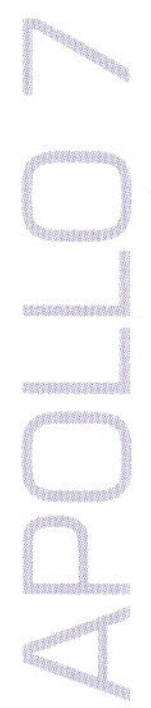

A crane lowers the Apollo 7 IU at the pad on April 17.

Technicians on both sides of the S-IVB guide the unit toward mating.

On April 17, BP-30 and its SLA are ready to be raised at LC-34.

BP-30 is lifted from its transporter.

BP-30 heads for the mating with the IU.

The top edge of the S-IVB shows discoloration from a small accidental oxidizer (nitrogen tetroxide) spill, which happened on April 21 during a fueling test of the CM's RCS. The chemical was flushed away with 400 gallons of water, which ran into the IU.

A crane using a spiderlike handling fixture is ready to raise the forward skirt of the S-IVB for inspection after the SM, SLA, and IU were destacked following the spill. Some damaged electronics in the IU were replaced. The domed top of the S-IVB's liquid hydrogen tank is visible. The accident delays Apollo 7's launch by several weeks.

CHAPTER 4

May–June 1968

Apollo 7's SLA is moved onto a scissor-lift trailer from the Super Guppy at the Cape Kennedy AFS Skid Strip on May 2 after arriving from NAR in Downey, California. Used for Apollo I, it had been returned to NAR in Tulsa, Oklahoma, for refurbishment.

The wrapped SLA and a crate of hardware are ready to be towed to the KSC Industrial Area.

The SLA nears the MSOB.

NAR technicians prepare to unload the SLA in the MSOB's high bay on May 2.

A bridge crane raises the SLA toward a vertical position just before its wrapping is removed.

Launch vehicle test conductor Jim Pugh and chief test supervisor Don Phillips discuss the Apollo 7 processing flow chart in the MSOB on May 6.

The S-IVB's forward skirt is parked on its trailer at the base of the LC-34 service structure on May 9. It had been removed for cleaning after the April 21 oxidizer spill.

Workers attach a crane to the S-IVB forward skirt.

On May 9, the skirt is raised into position for mating.

An IBM technician peers through the opening for the IU's umbilical door on May 11, which provides access for ground test equipment. The IU had also been removed after the spill to clean or replace several cables, antennas, and electrical connectors. The unit is hung with red "Remove Before Flight" streamers.

The IU is ready to be raised.

Initially steadied by taglines, the IU is slowly lifted.

The IU continues up. The service structure's control cab is at lower right.

IBM technicians work inside the IU after its mating with the S-IVB forward skirt, as an NAR worker watches from the Level 7 work platform at upper right.

On May 11, Eisele's suit ventilator will be disconnected before he enters the Apollo 7 CM for a Crew Compartment Fit and Function test (known as the C2F2) in an NAR clean room in Downey.

Cunningham prepares to join Eisele for the daylong session to make sure everything in the crew compartment works as expected. The astronauts wear cloth covers to protect their clear polycarbonate helmets.

Cunningham waits to ingress. Seated at right is Leo Krupp, NAR chief Apollo research pilot. Eisele's first name is misspelled on the welcome sign.

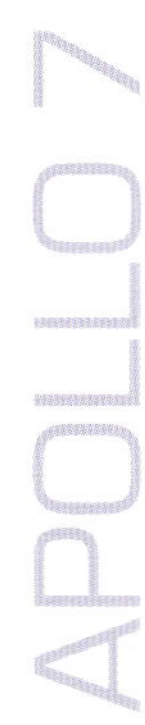

Schirra is ready to egress. The CM's exterior and hatch opening are protected by well-worn covers.

Schirra steps out.

Left to right: Cunningham, Schirra, and Eisele in front of their CM after the test

On May 17, NAR workers prepare to unload SM-101 for Apollo 7 from the Pregnant Guppy at the Cape Kennedy Skid Strip after it arrives from Downey.

In the MSOB, the top cover of the SM's shipping container is lifted.

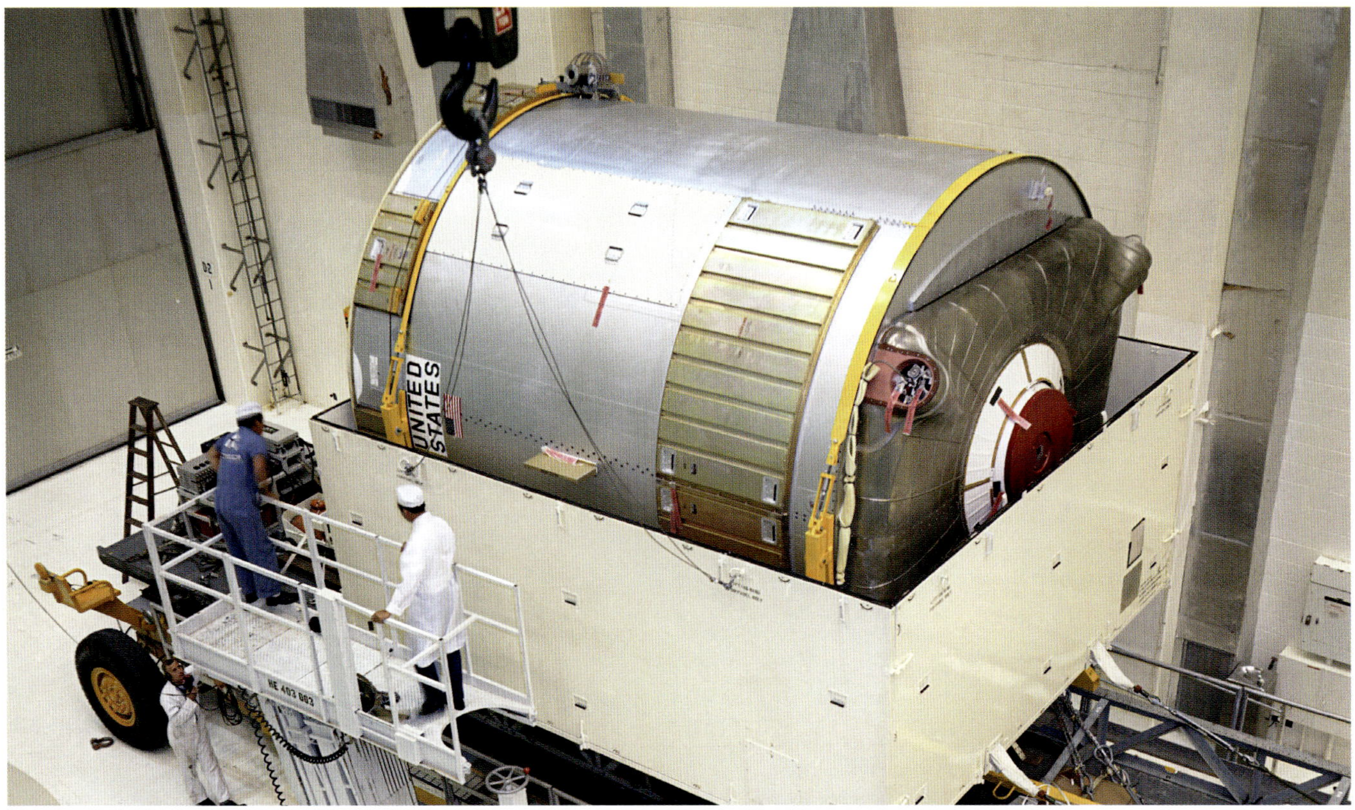

The nearly 13-foot-diameter SM is revealed.

On May 22, SM-101 is attached to a bridge crane to be moved into altitude chamber L.

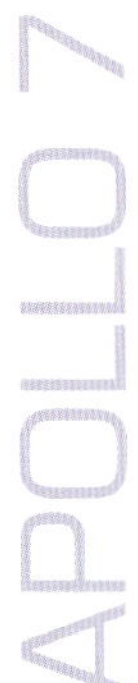

The crane slowly lowers the SM into the chamber.

The SM's umbilical connection to the CM is at left.

The two KSC altitude chambers were designed and manufactured by the Stokes Equipment Division of the Pennsalt Chemical Corp. of Philadelphia, Pennsylvania, and were installed in 1965 in the MSOB's high bay. All CSM tests used chamber L, a mirror twin of chamber R used for LM tests. Both chambers were about 33 feet in diameter and 59 feet high, with walls of 0.5-inch-thick stainless steel. They were first used in late 1965 to test CSM-009 for the AS-201 mission, but they were not man-rated until 1966. Although capable of simulating an altitude of 250,000 feet above sea level, the altitude tests on Apollo spacecraft were conducted at a simulated altitude of 210,000 feet.

The SM is almost seated in the chamber. Six hinged work platforms have been raised out of the way.

Technicians carefully guide the SM down. A red-and-white circular cover protects the engine thrust mount assembly where its Service Propulsion System (SPS) nozzle would be attached.

NAR and NASA technicians confer with the SM secured above them in the chamber.

Left to right: Eisele, Schirra, and Cunningham are seen through the hatch from inside a CM mockup at KSC on May 22. Couch struts are on the left and right. This mockup in the Flight Crew Training Building is used for emergency-egress training.

The crewmen strike a praying pose outside the CM mockup. The mission insignia on their suits are stickers, since nonflammable Beta cloth versions have not yet been sewn on.

Left to right: Schirra, Cunningham, and Eisele feign fear as they look over the side of a replica swing arm section that is attached to the new, larger white room mockup in the Flight Crew Training Building. It includes new safety features recommended by Apollo I investigators.

On May 30, Apollo 7's CM (CM-101) is unloaded from the Pregnant Guppy onto a scissor-lift transporter at the Cape Kennedy Skid Strip after arriving from NAR in Downey.

The CM protective cover bears banners signed by NAR workers.

CM-101 is unloaded from the Pregnant Guppy.

The CM, its environmental control unit, and a crate of hardware are lowered on the lift.

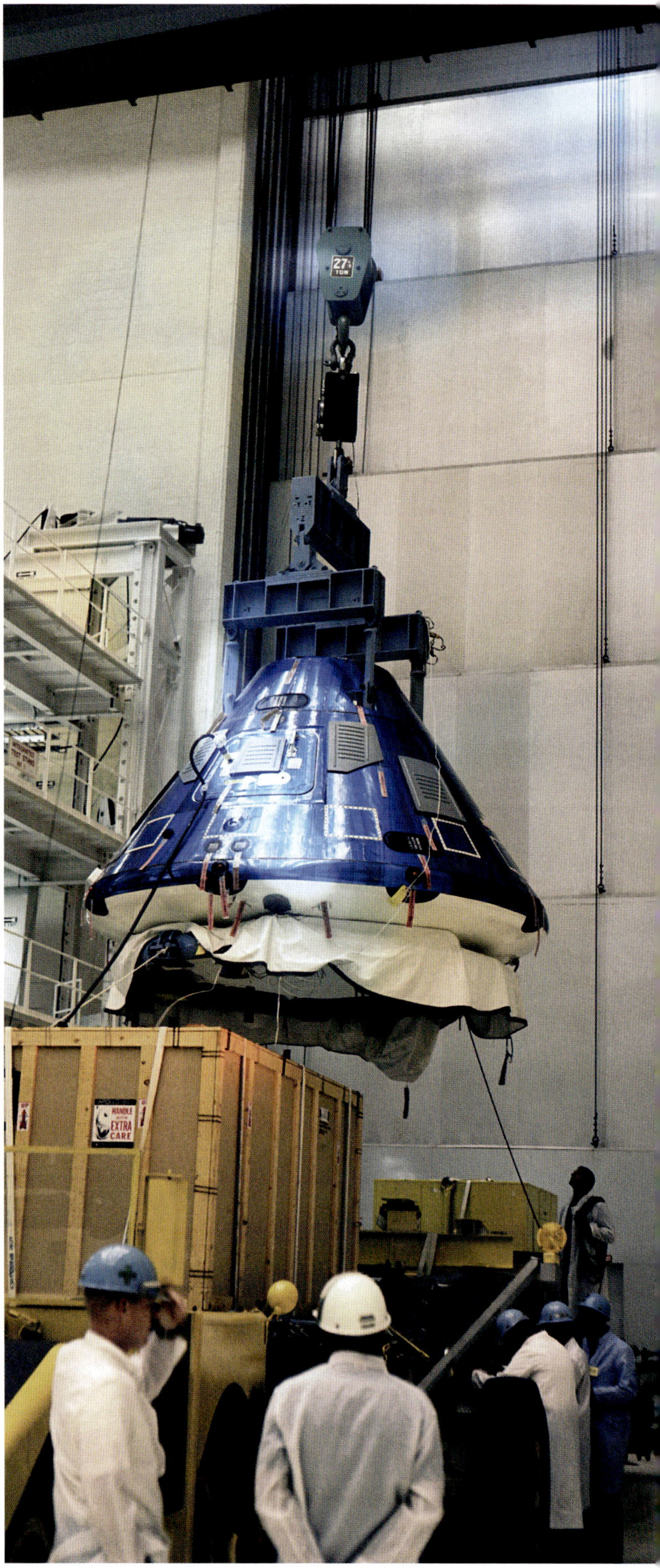

Technicians prepare to remove the CM's cover inside the MSOB.

The CM is raised from the transporter for transfer to a test stand. It arrives from NAR wrapped in blue plastic tape, which is not removed until several days before launch, when the Boost Protective Cover (BPC) is installed at the pad. The tape protects the underlying aluminized Kapton tape covering the CM to provide thermal control.

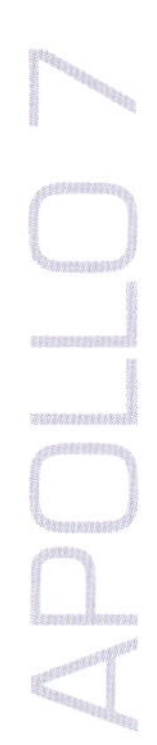

The CM is lowered into the test stand, with its yellow umbilical cover still in place.

The next day, Schirra (*left*) urges NASA and contractor personnel in the KSC training auditorium to work together to put the finishing touches on the CSM. "We're quite pleased with the spacecraft so far," he says. "We feel it's almost ready to fly." *Seated, left to right*: Walt Kapryan, deputy launch operations director; Buzz Hello, general manager of NAR's Space Division-Launch Operations; and Cunningham.

Dick Dunphy, who heads the RCA team that developed a 4.5-pound, black-and-white vidicon TV camera for the crew, looks through the viewfinder at MSC in an RCA photo released May 20. An identical camera would provide live video from Apollo 8 in lunar orbit in December.

RCA engineer Al Bishop, with the VAB reflected in his sunglasses at LC-39, holds one of the five cameras that was refurbished for the mission from the six first built for Block I use.

Dunphy demonstrates the RCA camera aboard a CM mockup.

On June 11, a bridge crane moves CM-101 toward altitude chamber L in the MSOB for mating with the SM.

The CM is ready to be lifted into the chamber.

Technicians mate the CM with the SM. They would next be separated for Apollo 7's reentry.

In June, NASA conducts 2TV-1, a manned vacuum chamber test at MSC in Houston, to certify a Block II CSM for spaceflight that incorporates changes made after Apollo 1.

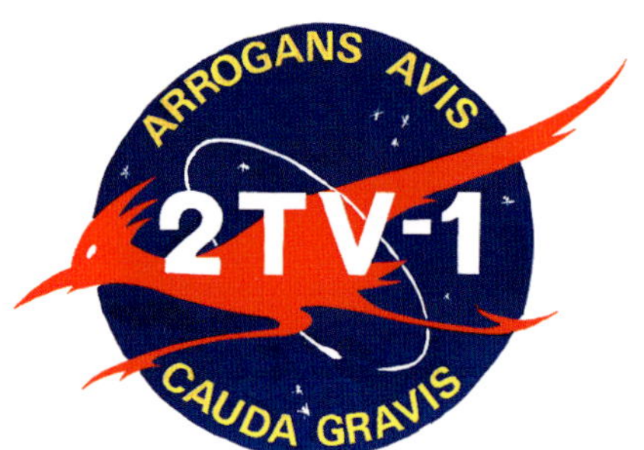

The emblem for the test, which the crewmen wear on their space suits and garments, is a spoof of the NASA insignia, with a roadrunner and the Latin phrase "Arrogans Avis Cauda Gravis" ("The Proud Bird with the Heavy Tail").

Workers secure 2TV-1 (CSM-098) in chamber A of the Space Environment Simulation Laboratory in Building 32 on May 22. The spacecraft, delivered from NAR to MSC in April, had been built side by side with CM-101 to duplicate Apollo 7's configuration. A rotating platform could simulate the spacecraft's "barbecue" mode in cislunar space to even out surface temperatures. The chamber could expose the CSM to temperatures up to 260 degrees F, using carbon-arc lamps to simulate solar radiation, and as low as minus 230 degrees F by circulating supercold liquid through the inner wall of the chamber.

Commander Joe Kerwin (*left*) and CM pilot Vance Brand wear A6L space suits aboard 2TV-1 on June 16, with LM pilot Joe Engle out of frame at right. They spend nearly eight days in the CM activating and checking out systems, operating guidance and navigation equipment, and simulating SPS engine firings.

Kerwin, the only astronaut physician, aboard 2TV-1. "We felt like we were breaking some ground and doing some necessary testing to enable the flight of Apollo 7," he said.

Engle (*left*) and Kerwin aboard 2TV-1. The crewmen remove their space suits a few hours into the test. "We didn't think of it as being quite such a monumental thing," said Engle. "We were kind of bored in there, actually. I think I pulled on my hunting and camping skills to living in a confined area."

Left to right: Kerwin, Brand, and Engle are happy to leave the chamber on the morning of June 24.

Left to right: Brand, Engle, and Kerwin dress as hippies later that day to visit MSC deputy director George Trimble "with our week's beards" to assure him that the spacecraft is ready for the prime crew. Recommendations produce twelve hardware design and thirteen crew procedure changes for Apollo 7.

On June 21, a new Astronaut Transfer Van, a modified Clark Cortez Motorhome, is delivered to KSC. The Apollo 1 crew was the last to use its predecessor, a 1965 International Metro M-1200 van, which also transported the final four Gemini crews. The astronauts and suit technicians would board the van at the MSOB for the ride to LC-34 (Apollo 7) or LC-39. It would remain in use through STS-6 in 1983.

View of rear right side of the van. The open hatch shows the empty propane tank compartment, which would instead house photo and movie cameras to record the crew leaving from the rear door at the pad. A window for them would be added below the red reflector.

The instrument panel includes two toggle switches for crew compartment lights (*center*).

An air conditioner and fire extinguisher are mounted next to the driver on the right side.

The service structure (*right*) has been partially rolled away from the umbilical tower (*left*) and the AS-205 Saturn IB at LC-34 in this view looking south-southeast on June 27. Two gray flame deflectors are at lower right, with the blockhouse behind them. A boilerplate SLA has been mated to the IU during swing arm testing.

The blue spotlights in the foreground will eventually be spaced around the booster for night illumination. This view looks north-northwest.

The service structure has been rolled to its parked position, with the Atlantic Ocean in the distance, on June 28 in this northeast view during photo site tests. The cable run to the blockhouse is at lower right. LC-37 is at left.

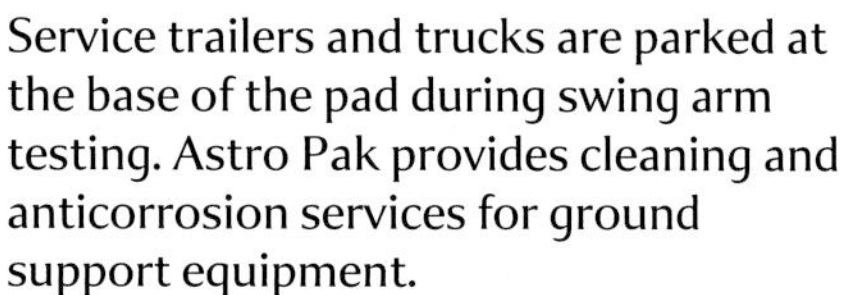

Service trailers and trucks are parked at the base of the pad during swing arm testing. Astro Pak provides cleaning and anticorrosion services for ground support equipment.

Pipes for the pad water deluge system are in the foreground in this view looking up at the umbilical tower during swing arm testing. The three lower swing arms contain fuel and electrical lines; the white room is at the end of the crew access swing arm, with the pad crane above it.

AS-205 during swing arm testing

The boilerplate SLA is lifted after demating on June 28.

The SLA is lowered.

CHAPTER 5

July–August 1968

A worker on the LC-34 umbilical tower near the crane monitors a water deluge test on July 1. A Firex Water System supplies industrial water to four spray nozzles on each swing arm, available both for fire suppression and to wash away any spilled fuels.

Technicians race across the crew access arm underneath sprinklers toward the white room during the test.

Water pours into Level 8 of the service structure, with the covered white room at center, and green tarps at bottom where the CM would be. The boilerplate SLA has been removed, so only the first and second stages and the IU are below.

On the afternoon of July 4, technicians help Eisele ease into the CM for emergency pad evacuation training in altitude chamber L in the MSOB. Both the prime and backup crews had practiced it while wearing smocks that morning. The drills have been implemented as part of the post–Apollo I safety changes.

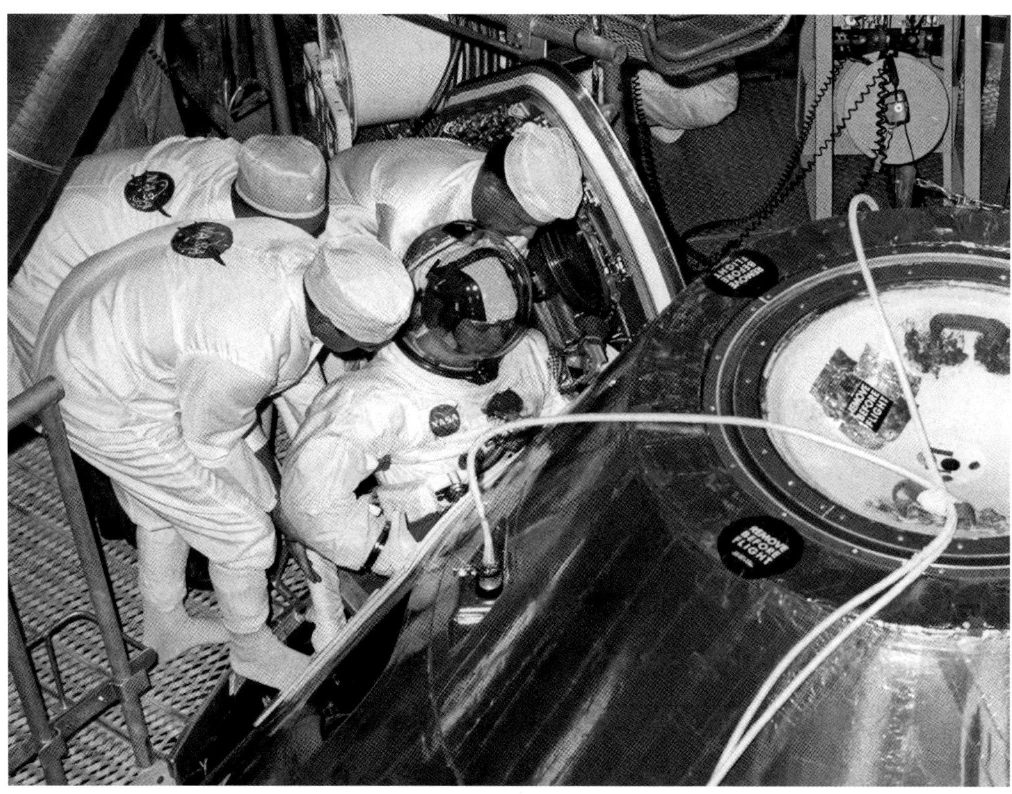

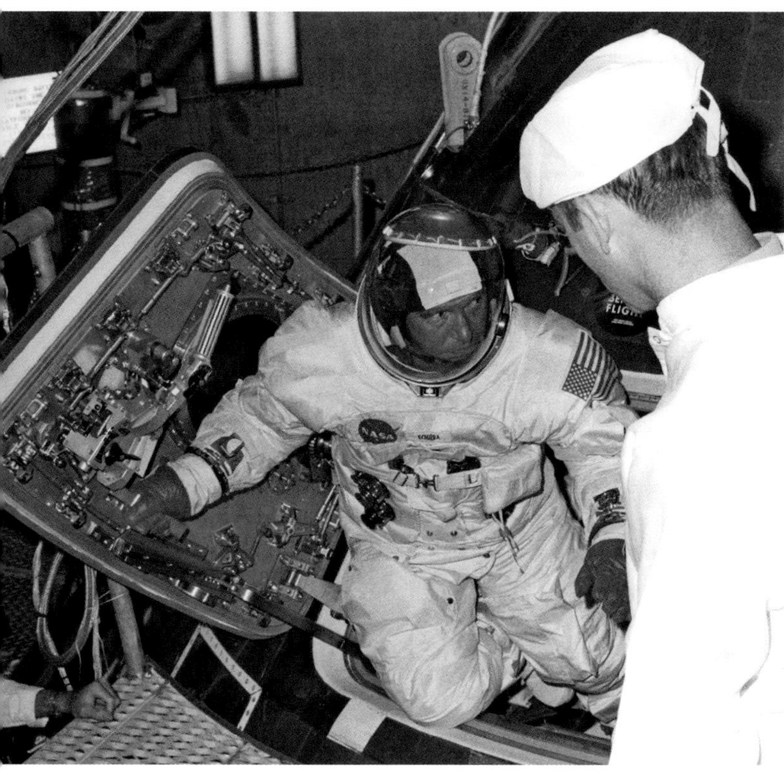

Schirra climbs out of the CM after the test, which has the new unified hatch. The astronauts practice egressing with and without assistance.

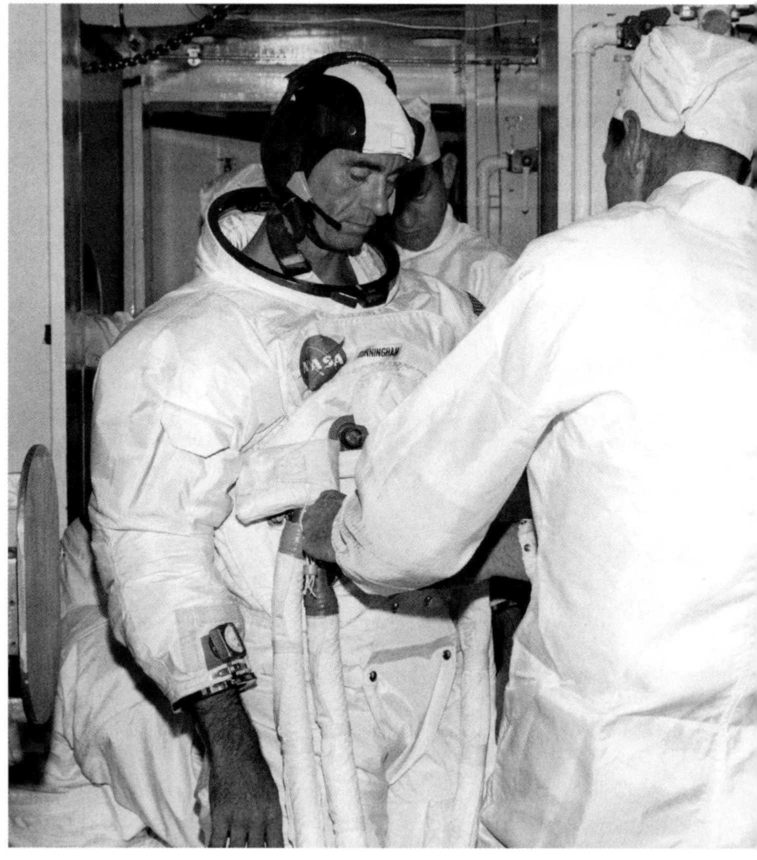

NASA suit technician Clyde Teague (*right*) works with Cunningham after the test.

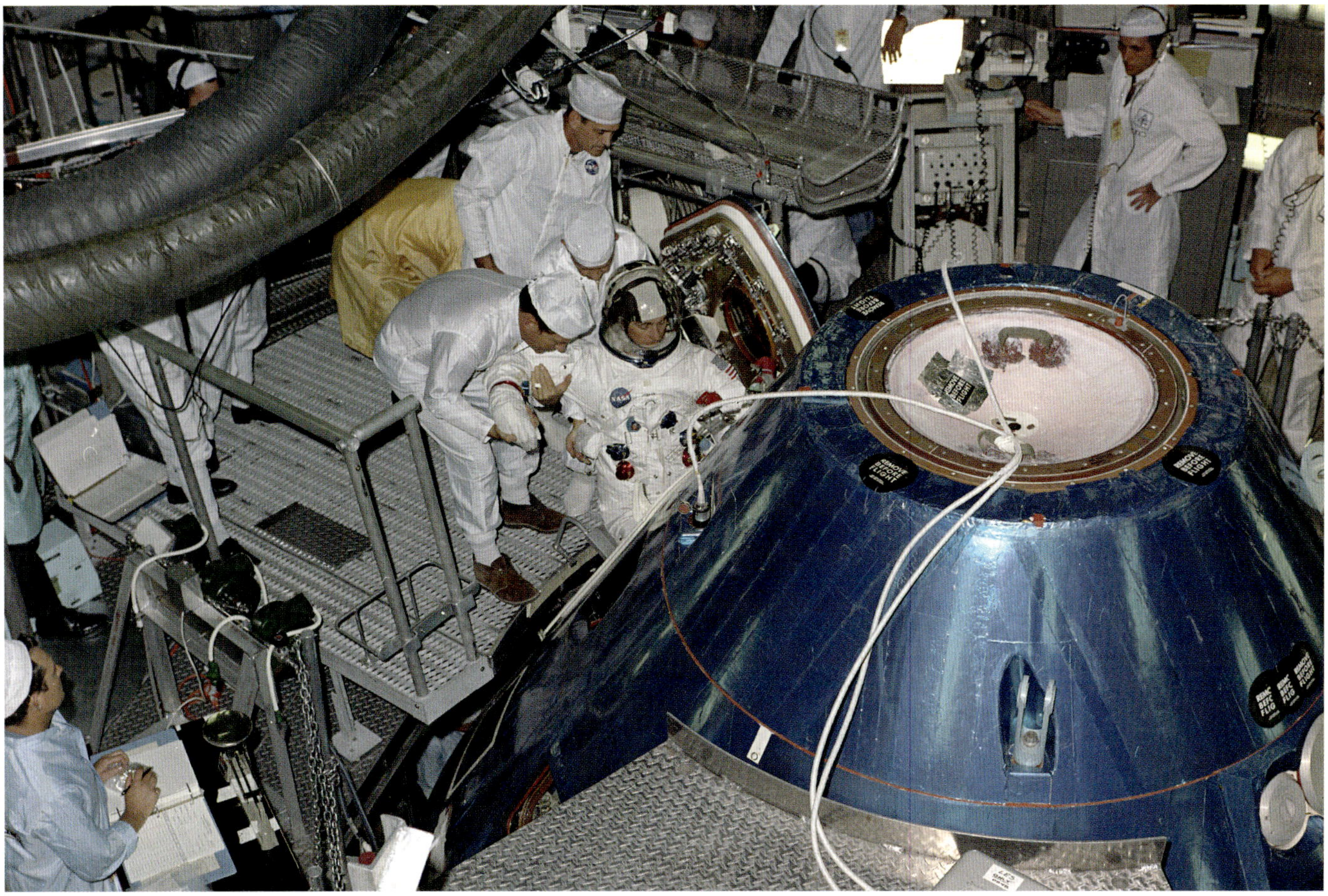

The next day, the backup crew repeats the egress training while suited. Backup LM pilot Gene Cernan carefully slides into the CM second. Three aluminum stretchers are nearby at the top of the photo.

Backup CM pilot John Young, occupying the center couch, will be the last to ingress.

Two NAR techs wearing oxygen masks open the CM hatch as part of the emergency-egress training.

On the morning of July 8, technicians make final checks for the prime crew's first true altitude chamber run before sealing the hatch.

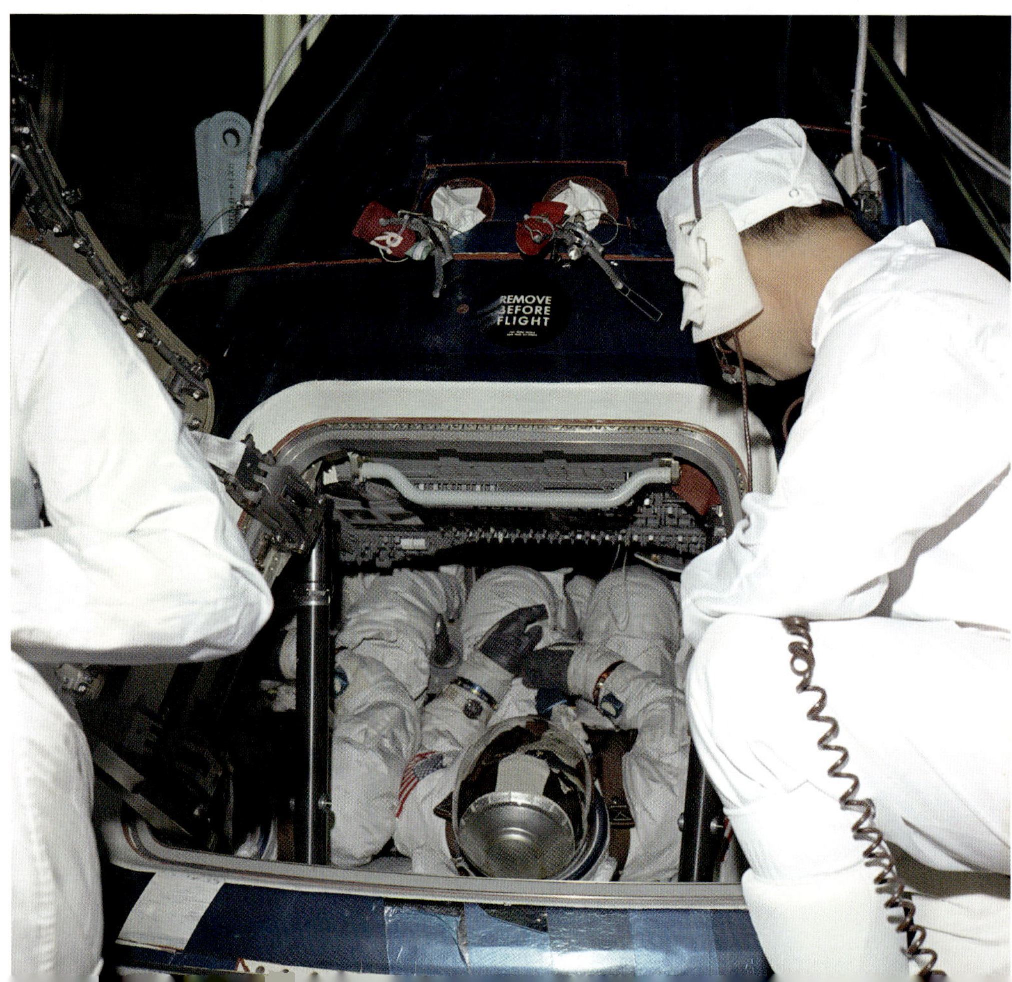

A TV camera (one of the five supplied by RCA for Block II use) is included on board.

NAR pad leader Guenter Wendt peers through the hatch window during the eight-hour chamber test.

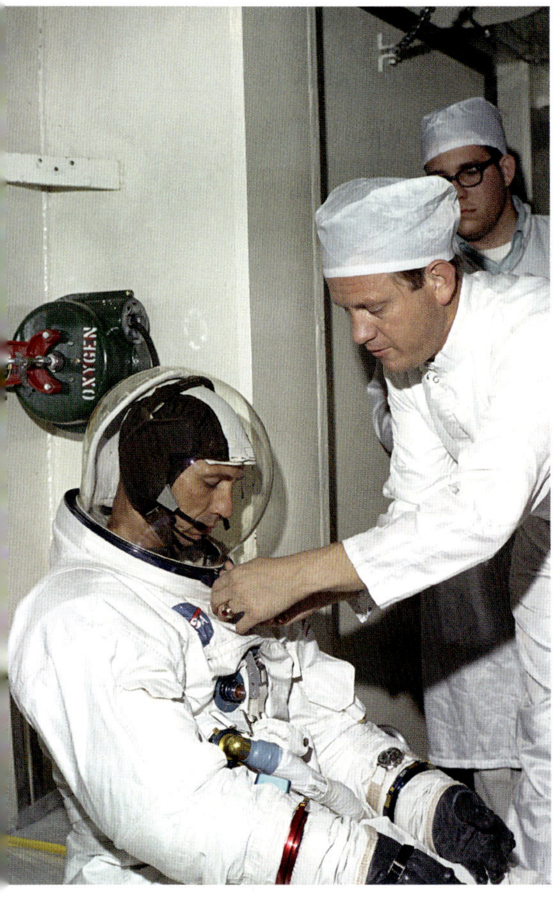

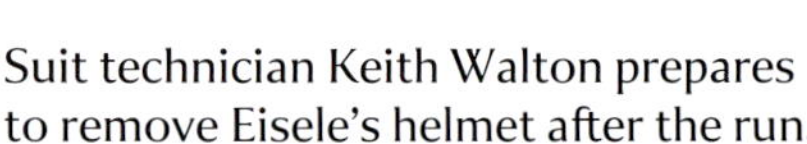
Suit technician Keith Walton prepares to remove Eisele's helmet after the run.

Cunningham leaves the chamber, with Young at right.

This docking target, photographed in the MSOB on July 22, would be mounted inside the SLA for use during a rendezvous test on the first day of the mission. It duplicates those on the LM for docking during lunar missions. The target was designed by Royal Australian Air Force ophthalmologist Dr. John Colvin.

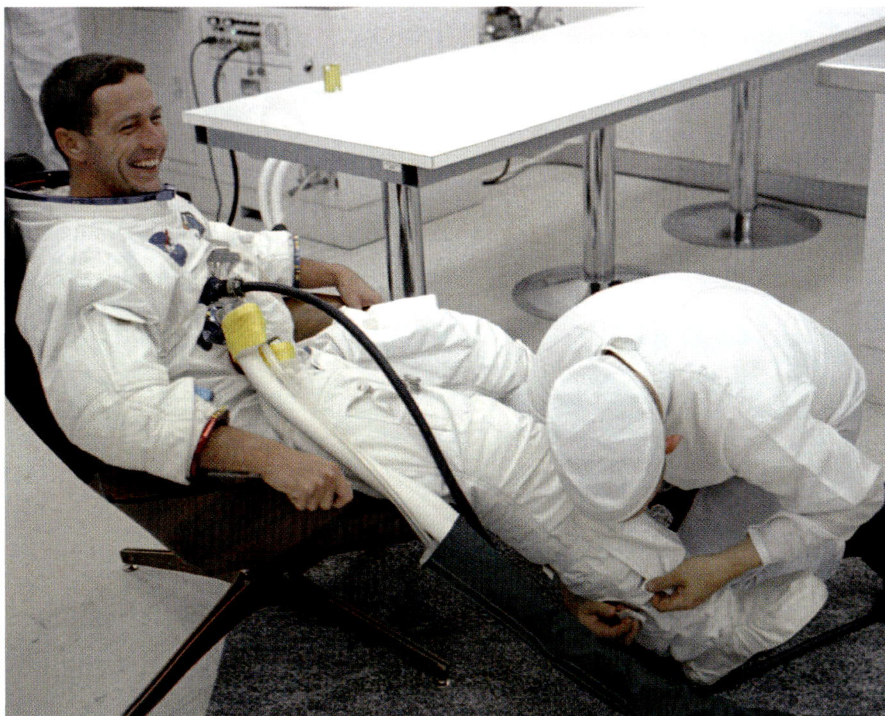

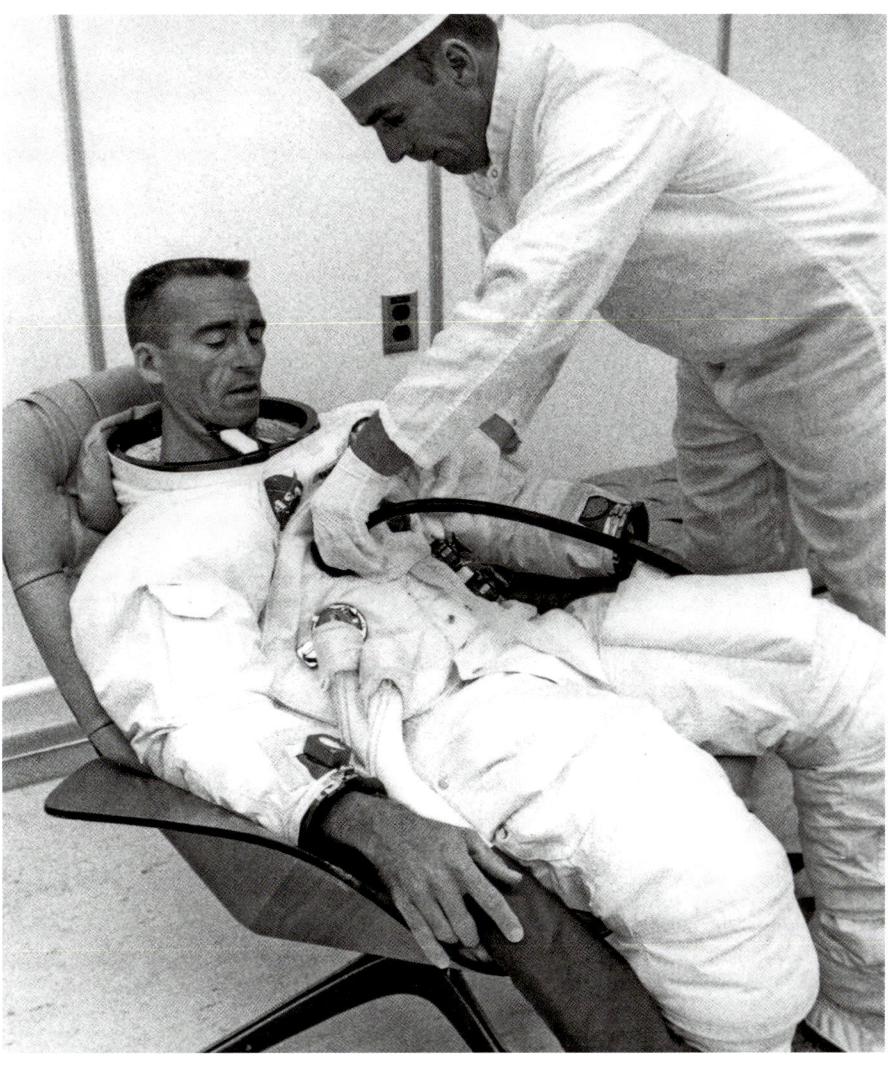

On the morning of July 26, Schirra relaxes in the MSOB suit room before the prime crew's final altitude chamber run. Two days earlier, the CM had completed a twelve-hour unmanned chamber test.

Eisele's suit cuff is adjusted. The suit room maintains two flight suits and one training suit for each prime crewman. Each of the backup astronauts has one flight and one training suit.

Suit tech Jim Lewis attaches the electrical connector to Cunningham's suit.

Schirra reaches out to the camera in altitude chamber L, with Stafford at right.

Eisele (*right*) talks with Dr. Alan Harter, KSC staff physician, who monitors the crewmen during chamber runs.

Cunningham waits to enter the CM for what would be a more than nine-hour test.

Schirra is ready to ingress.

Technicians help Schirra ease into the CM. "The spacecraft, the test team, and we, the crew, had a good run," Schirra says later. "Our confidence in the spacecraft was so high that we were able to vent the cabin and remain for several hours at high altitude in our pressure suits, relying solely on the spacecraft systems for support."

Eisele hands his ventilator to a technician before ingressing just before noon. The crew would spend nine hours inside the spacecraft, most of the time at a simulated altitude of 226,000 feet, and, for the first time since Apollo I, in a pure oxygen environment.

On July 29, backup commander Stafford waits to enter the CM before his crew repeats the chamber test, this time at altitude.

Suit techs prepare backup LM pilot Cernan for ingress.

Backup CM pilot John Young will be the last to ingress. The NAR technician at right, still wearing an NAA patch, serves as "pad leader" (identified by the yellow smock).

Apollo 7's launch vehicle (SA-205) undergoes final pad fit and function tests at LC-34 on July 30, this time with a boilerplate CSM in place.

The Saturn IB with its umbilical tower on July 30

Left to right: Backup crew members Cernan, Young, and Stafford take part in recovery training in the Water Immersion Facility in Building 5 at MSC on August 2.

On August 5, technicians prepare to raise the Apollo 7 CSM from the altitude chamber.

The spacecraft is lifted out of the chamber.

The bridge crane moves the CSM toward a workstand.

The CSM is in place on the stand, with the SPS nozzle extension ready for attachment below.

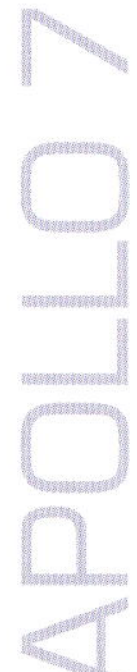

NAR technicians prepare to bolt the extension to the SPS engine thrust mount assembly. Its gimbal ring allows the nozzle to move about 7 degrees up or down (pitch) or left or right (yaw), driven by a pair of electromechanical actuators.

The CSM is moved into position above the SLA in an integrated test stand.

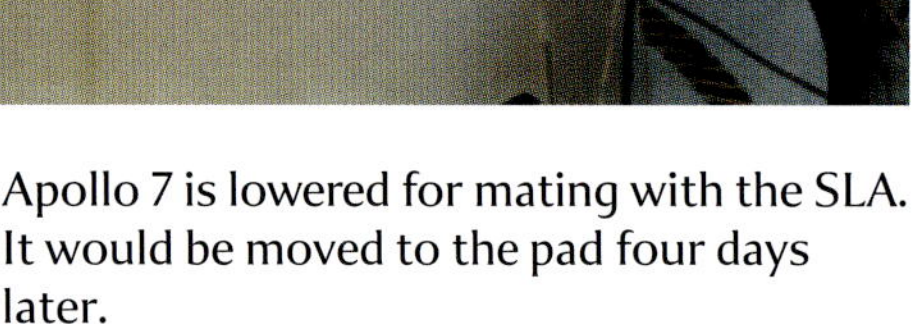

Apollo 7 is lowered for mating with the SLA. It would be moved to the pad four days later.

On August 5, Schirra (*left*) and Cunningham emerge onto the deck of MV *Retriever* after getting suited for recovery training in the Gulf of Mexico off Galveston, Texas.

Eisele waits with his suit ventilator. In the background is *Duchess*, a yacht owned by La Porte, Texas, businessman Paul Barkley and loaned as an observation boat for journalists covering the exercise.

Cunningham, Eisele, and Schirra pose with their suit ventilators on the deck of *Retriever* before entering BP-1101A.

BP-1101A was an aluminum boilerplate built to test the CM's flotation characteristics. It was fabricated at Kelly AFB in San Antonio, Texas, and was delivered to MSC as BP-1101 in early 1965 in a Block I configuration. It was returned to Kelly for modifications to Block II as BP-1101A in late 1965. The boilerplate used internal ballast and exterior dummy equipment to approximate equipment location in the crew compartment. It was used to test the Block I uprighting system and the flotation collar in MSC's Water Immersion Facility and in the Gulf of Mexico. The uprighting system was commanded by a hardwired box controlled by the test conductor from a life raft.

Schirra (*foreground*) and Cunningham, seen through a side window of the boilerplate on the ship's deck, are ready.

The boilerplate with astronauts aboard is lowered into the Gulf of Mexico.

The CM floats in its stable II (apex down) position as uprighting bags inflate to flip it over.

The next day, backup crew members (*left to right*) Cernan, Stafford, and Young are readied below deck to repeat the training.

Stafford prepares to enter the CM boilerplate.

Stafford joins Young (*left*) and Cernan in a life raft; Young has removed his pressure suit.

The Apollo 7 CSM and its SLA wait at the MSOB's open high bay door at 2:00 a.m. on August 9 for transport to LC-34.

The CSM arrives at LC-34 at dawn.

The CSM and SLA are raised at LC-34 on August 9.

The CSM will be moved horizontally over the booster.

Apollo 7 is lowered toward the Saturn IB with two pairs of hurricane doors retracted.

The Apollo 7 spacecraft is nearly mated with the Saturn IB's IU at 8:30 a.m.

The Launch Escape System (LES) arrives at LC-34 later in the day for installation atop the CM. The Lockheed Propulsion Co., northeast of Redlands, California, provides both launch escape pitch control motors of the LES, which used a solid polysulfide fuel. The Thiokol Chemical Corp. of Ogden, Utah, provides the tower jettison motor.

The new Astronaut Transfer Van, a modified Clark Cortez Motorhome, makes its debut on August 14, when it ferries the Apollo 7 crew about two blocks from the MSOB to the Flight Crew Training Building for emergency-egress training.

The van's propane tank compartment is now occupied by a 35 mm still camera (*left*) and a 16 mm motion picture camera to capture the astronauts exiting from the back door at the pad.

The van's interior includes seats for three crewmen at right, each with a storage box for their suit ventilators. Communication headsets are on the seats at left.

Van driver Steve Tatham of KSC Security points out the Apollo 7 mission emblem on the rear door as he holds it for Schirra. Following behind (*right to left*) are suit tech Keith Walton, Eisele, Cunningham, and Clyde Teague.

Eisele, Cunningham, and Schirra share a laugh before boarding.

Schirra is first into the van.

Cunningham follows.

Eisele boards the van with suit tech Keith Walton.

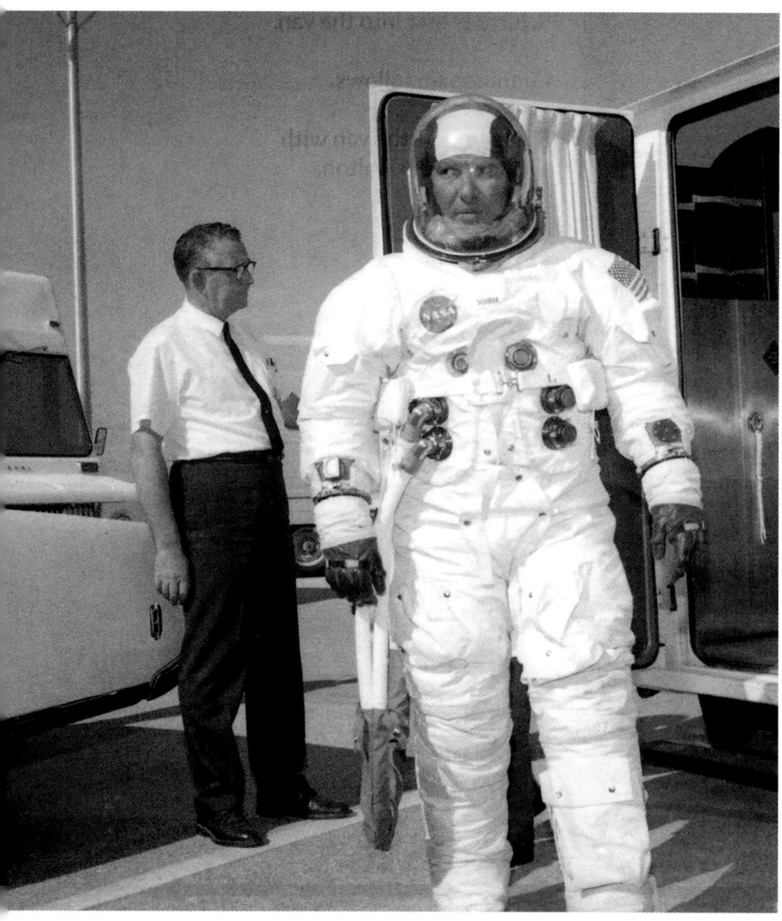

Schirra emerges from the van after arriving at the Flight Crew Training Building about two blocks away.

Cunningham is next out, followed by Jim Lewis.

Eisele and Walton follow them into the building for emergency-egress training.

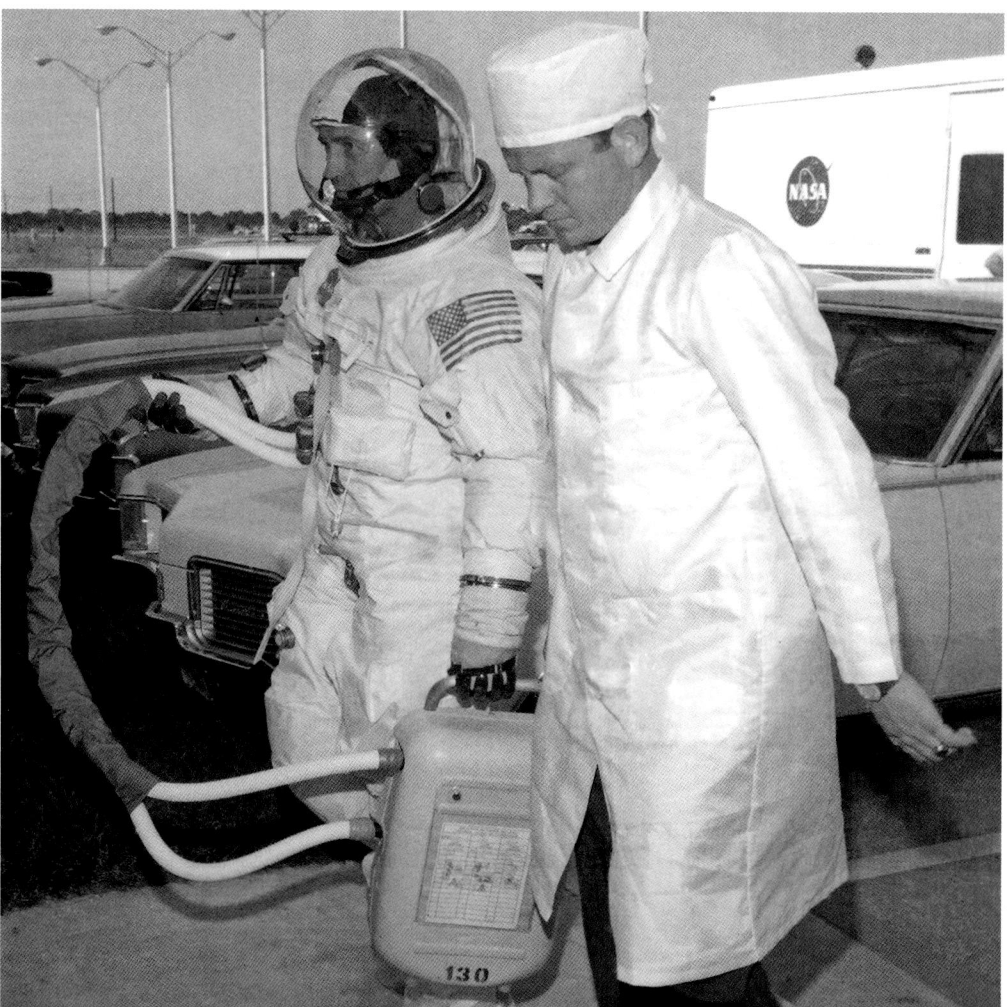

Schirra steps into the CM mockup to occupy the left couch.

Cunningham watches his commander ingress.

Eisele, who'll be last in, waits for his fellow crewmen to board.

On August 15, technicians at LC-34 prepare for two days of tests with a new slide-wire, an alternate escape route from the 224-foot level of the umbilical tower. It was designed by Chrysler.

Apollo 7 support crew members Bill Pogue (*sunglasses*) and Ron Evans wait at the end of the slide-wire before they don space suits to test the automatic braking system.

Pogue (*left*), with sling, and Evans are ready for testing after getting suited in a nearby trailer.

Evans and Pogue will be lifted by a Hi-Ranger aerial crane to the top of a temporary 36-foot scaffold (in front of the service structure on page 124) for the brief ride down.

Pogue begins his slide.

Pogue gets checked after his trip.

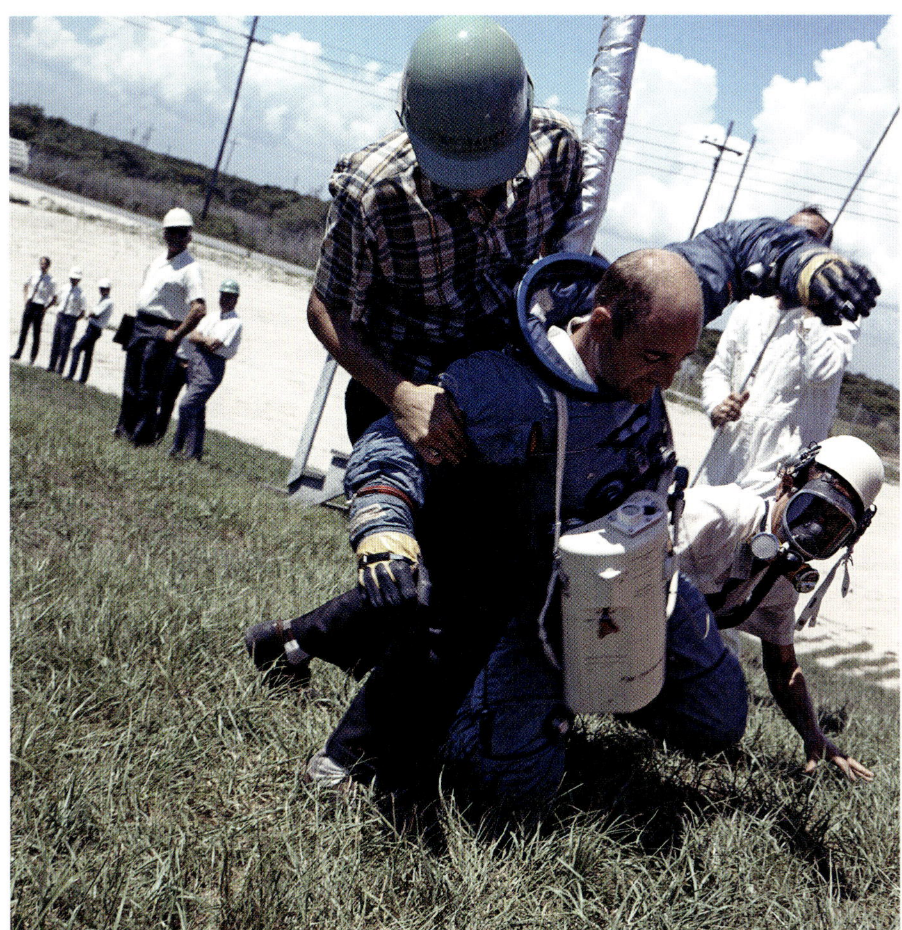

KSC safety personnel remove Evans from the slide-wire. He wears an emergency air supply around his neck, since an escaping crewman could have his helmet on.

Evans talks with NASA personnel after the testing.

The next day, NASA engineering manager Jim Ragusa is the first person to test the entire 1,200-foot-long slide-wire from the umbilical tower. "The ride was extremely smooth and vibration free, and for my run, it performed exactly as designed," he says. Planned tests of the full slide-wire with Pogue and Evans that afternoon are canceled after the brake system fails and a series of dummies have rough landings.

Eisele (*right*) discusses his suit with Cunningham in the Flight Crew Training Building on August 22 before emergency training.

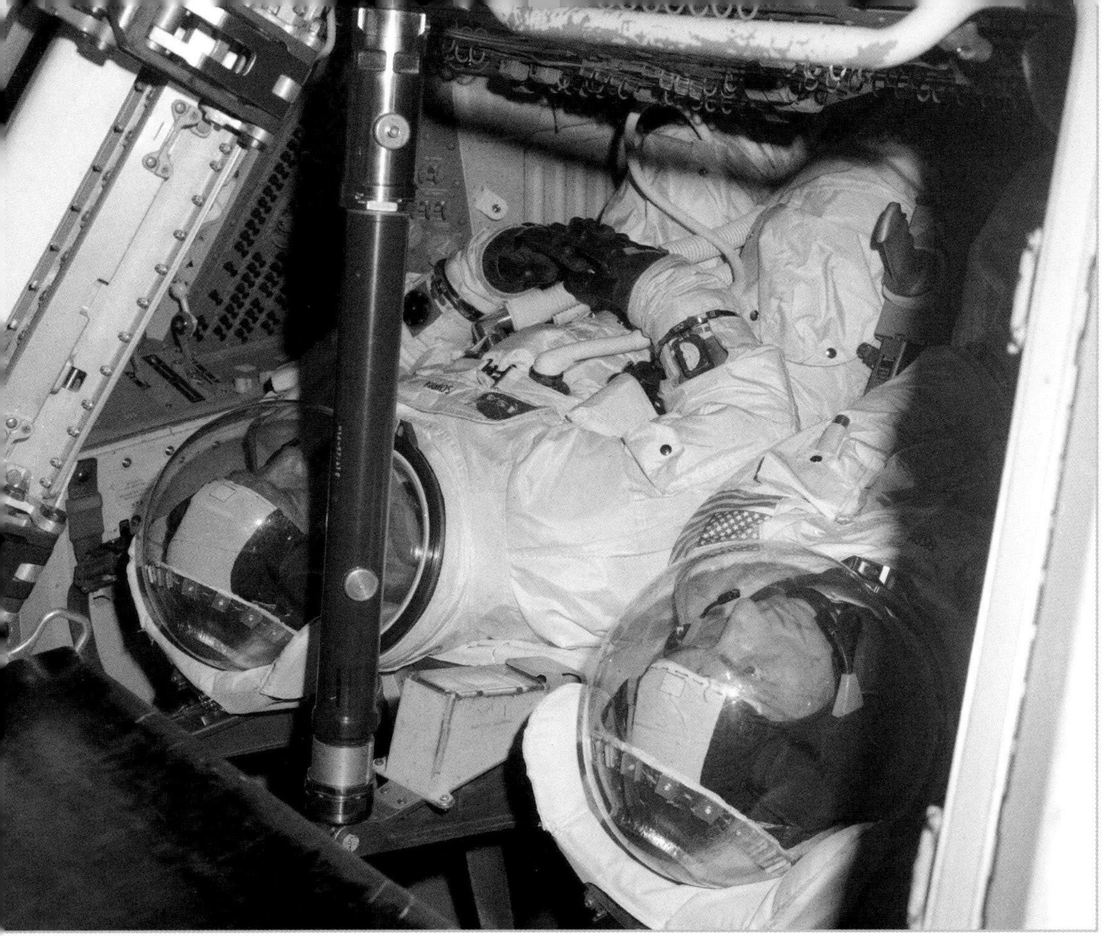

Schirra and Eisele in CM mockup no. 1, used only for ingress and egress training.

Eisele (*left*) and Cunningham inside the mockup

On August 27, Schirra enters the mockup LC-34 white room in the Flight Crew Training Building for an emergency-egress exercise.

Eisele enters the mockup, built at KSC. It duplicates the new larger white room designed to accommodate the unified hatch, and includes an exhaust fan, improved lighting, and new firefighting equipment.

Cunningham dons his gloves.

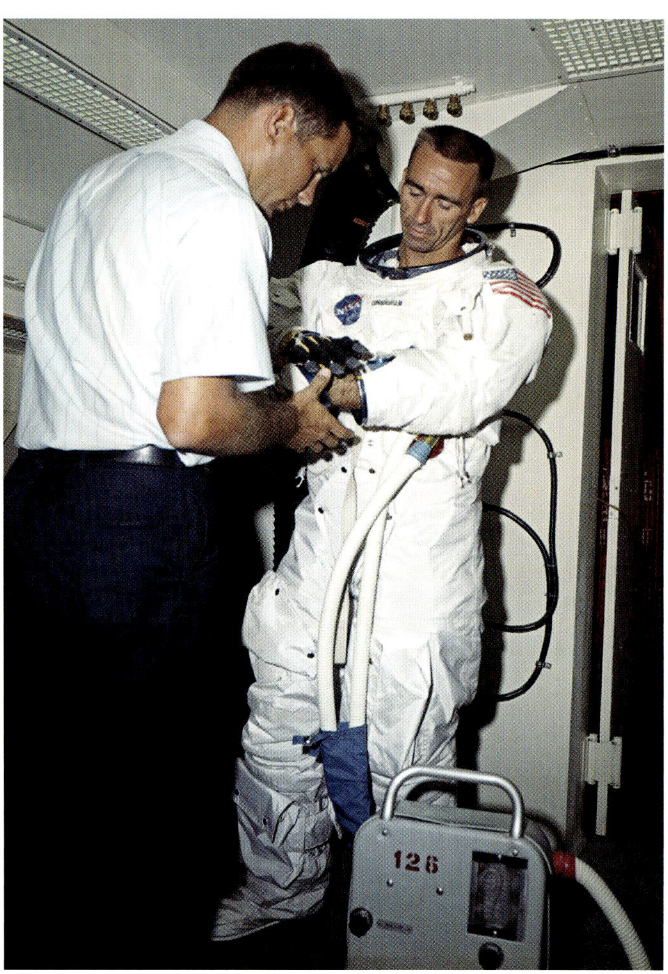

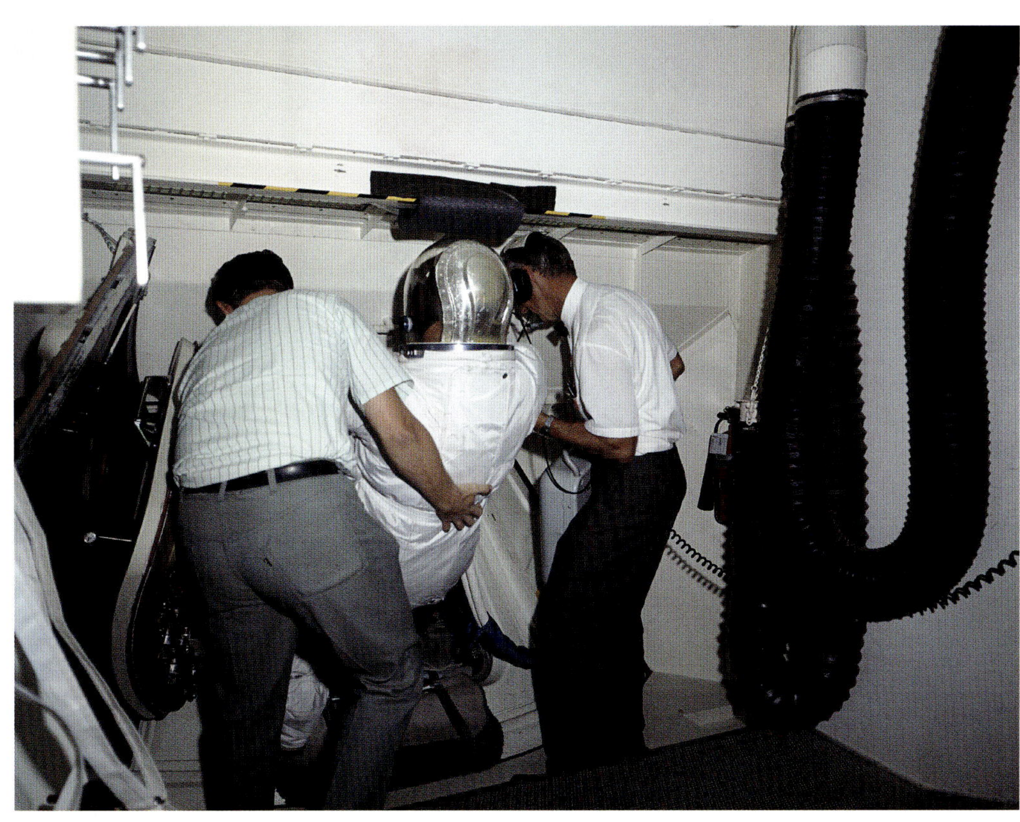

Pad leader Guenter Wendt (*right*) assists the crew into the mockup CM.

Eisele is on the simulated crew access arm after making an exit from the mockup on August 27. He is equipped with an emergency oxygen pack. At upper right are windows along a walkway for public tours conducted by KSC contractor TWA.

Cunningham (*center, talking with Wendt*), Schirra, and Eisele discuss the exercise with support personnel.

The next day, August 28, backup crew members Cernan (*left*) and Young enter the mockup white room for emergency-egress tests.

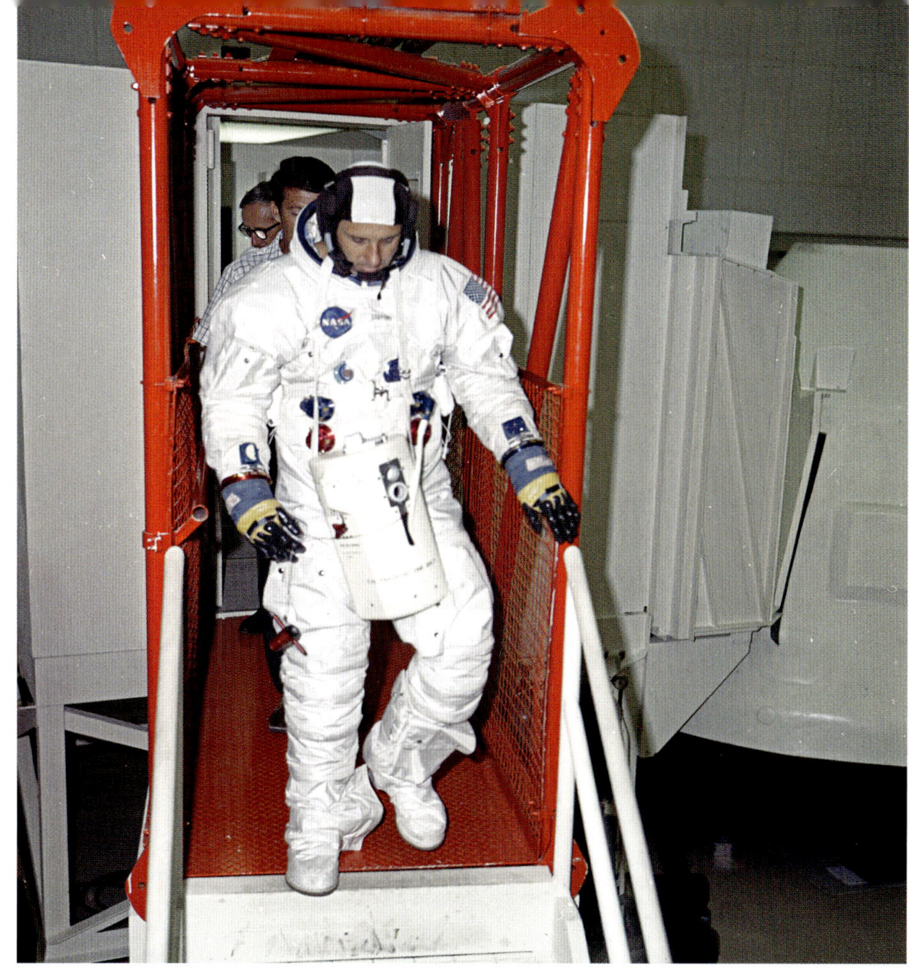

Backup commander Stafford steps off the mockup crew access arm after the exercise.

Schirra (*left*) talks with Stafford, Cernan, Young, Wendt (*glasses*), and NASA managers after the tests. Between Cernan and Wendt is Tex Ward of MSC's Flight Crew Support Division, chief of the Apollo egress-training program.

On August 30, Rene Rozalski (*left*), slide-wire test conductor for base contractor Pan American World Airways (PAA), and PAA LC-34 manager Evertt Crouse monitor the final tests, which see Crouse, Evans, and three safety managers successfully whiz down the wire ten seconds apart about 3:30 p.m. They reach peak speeds of almost 50 mph. Crouse is the final man down.

Evans is the third of the five to jump off the umbilical tower, reaching the ground in about eighty seconds.

Safety personnel prepare to release Evans at the arresting system at the end of the slide-wire. Camerman Jack Kenny films at right.

Ragusa (*left*) works with astronaut Stu Roosa as a Bendix worker pulls down the slide-wire (*at right*). Roosa, a former smoke jumper for the US Forest Service, took a short ride toward the backstop that cushions the landing.

Roosa (*center*) meets with Bendix Safety and Cape Fire Dept. representatives. At right is Elmer Brooks, PAA rescue team leader, who'd been the second to come down the slide-wire. Assigned to develop rescue procedures, Roosa had just been named to the Apollo 9 support crew and would test a similar slide-wire system for the Saturn V at LC-39A several months later.

CHAPTER 6

September 4–12, 1968

Schirra (*left*) holds his communications carrier (headset) in the LC-34 white room on September 4 as he and Cunningham prepare for a simulated countdown and launch.

Left to right: Cunningham, Schirra, and Eisele discuss white room equipment.

Schirra (*left*) and Cunningham adjust their chinstraps. The door to the crew access swing arm is behind Cunningham.

Cunningham enters the Apollo 7 CM for the first such exercise since the Apollo I fire.

NAR personnel monitor the astronauts during the mock countdown. “All is going well aboard the Apollo 7 spacecraft,” Schirra reports. Everyone—including the crew—wears socks.

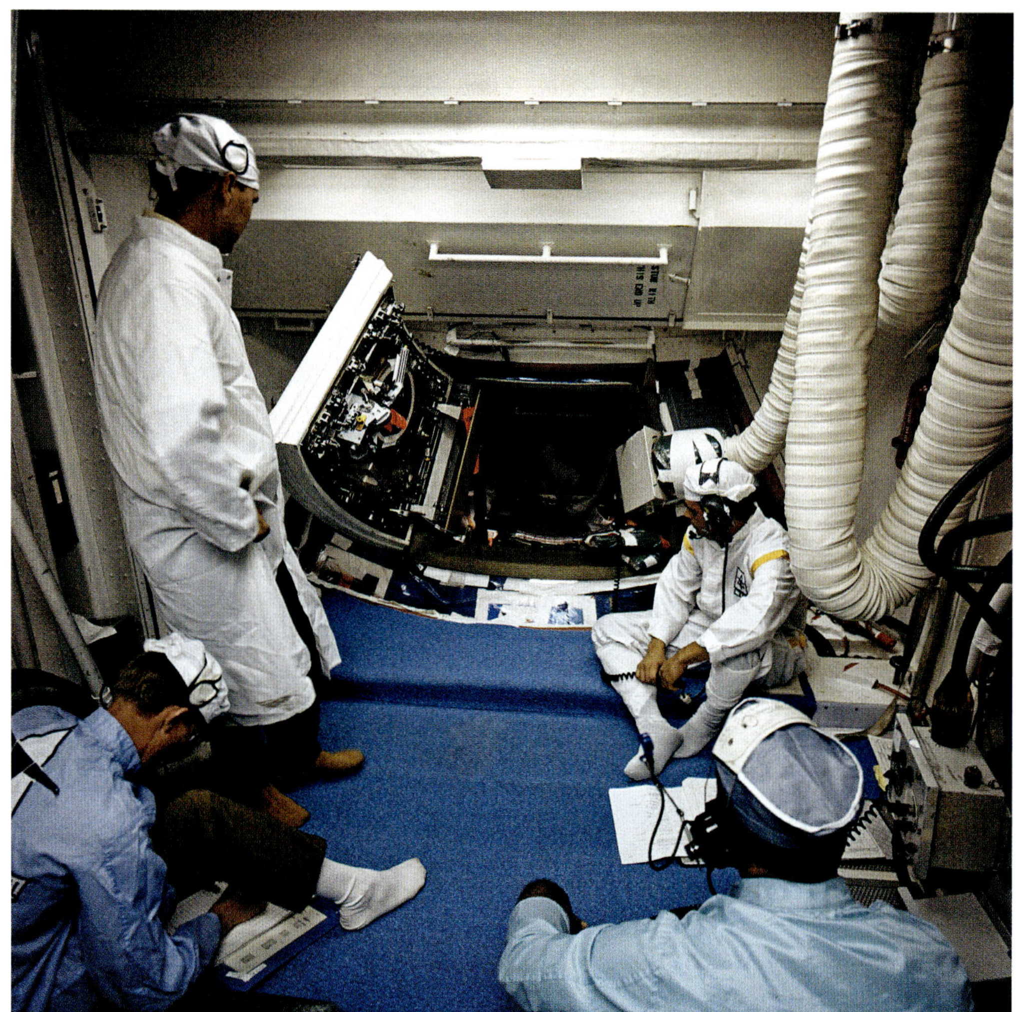

The new unified hatch remains open throughout the countdown, which ends with a successful “launch” at 4:26 p.m. EDT. A power glitch, however, forces the crew to repeat the test the next day.

On September 9, the astronauts participate in a series of suited emergency-egress drills at LC-34, including using the slide-wire. The backup crew goes through just the white room portion later that day.

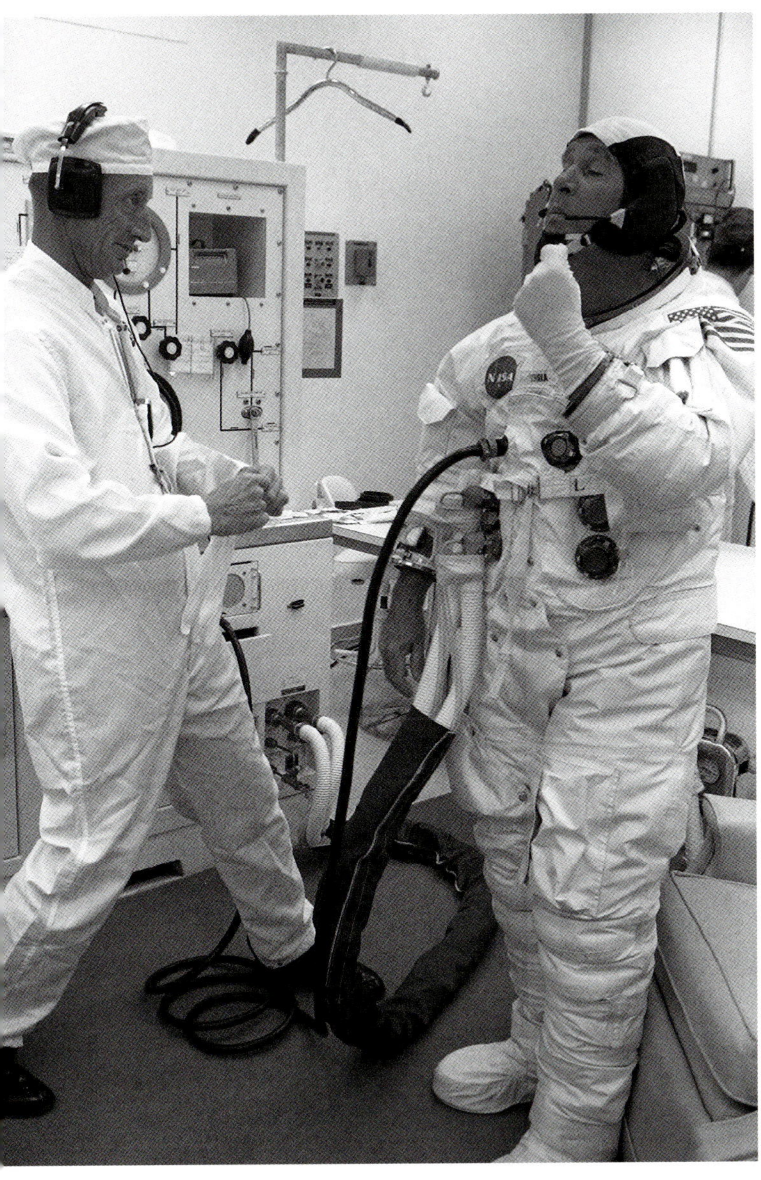

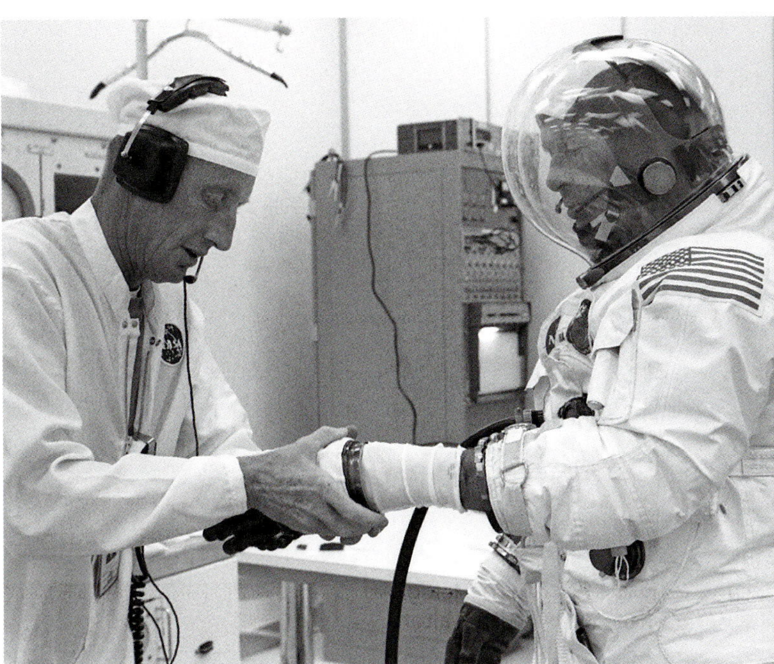

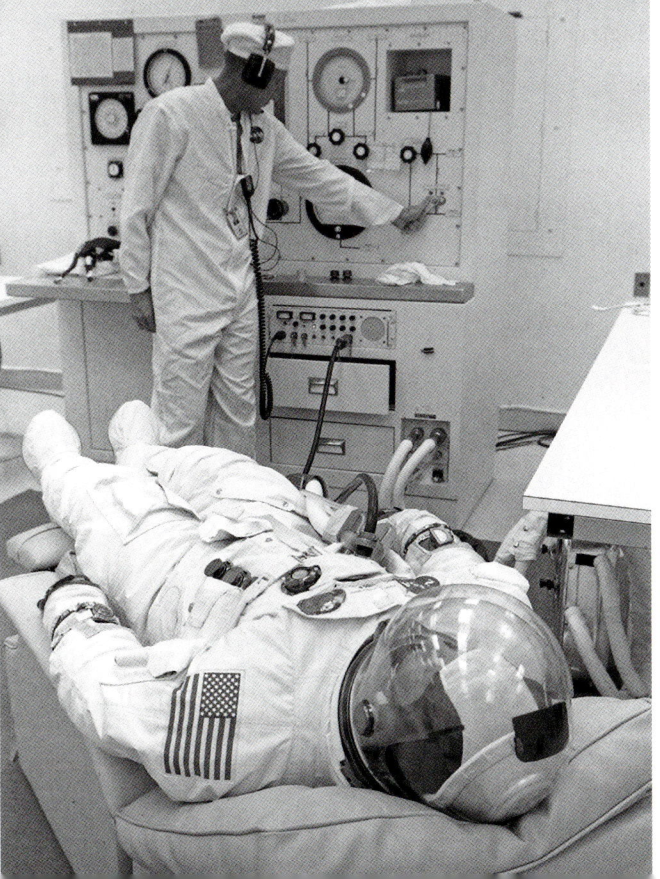

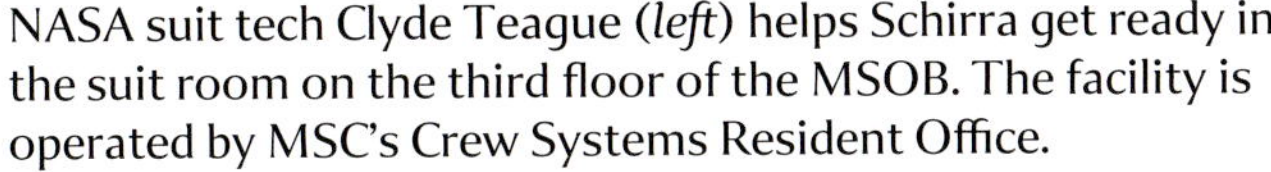

NASA suit tech Clyde Teague (*left*) helps Schirra get ready in the suit room on the third floor of the MSOB. The facility is operated by MSC's Crew Systems Resident Office.

Teague slides on Schirra's left glove.

Schirra relaxes in a recliner as Teague performs a suit leak check.

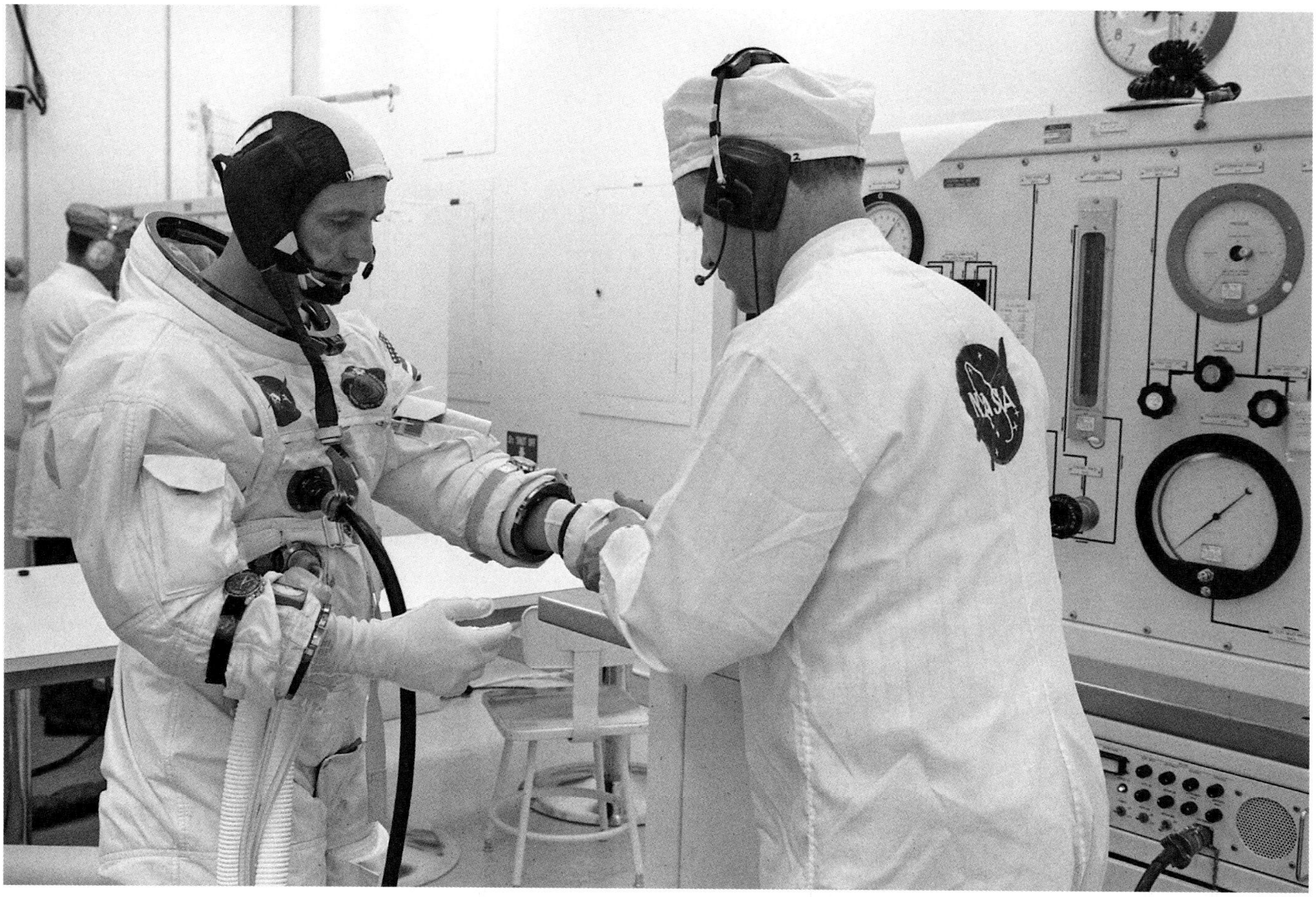

Eisele (*left*) has his left glove attached by Keith Walton.

Eisele is ready for his helmet.

Eisele checks his helmet's fit.

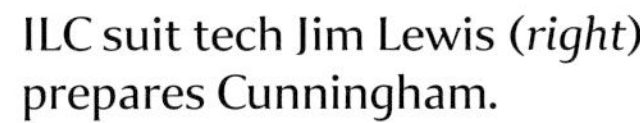

ILC suit tech Jim Lewis (*right*) prepares Cunningham.

Cunningham is ready for a suit leak check.

Cunningham relaxes before departing for the pad.

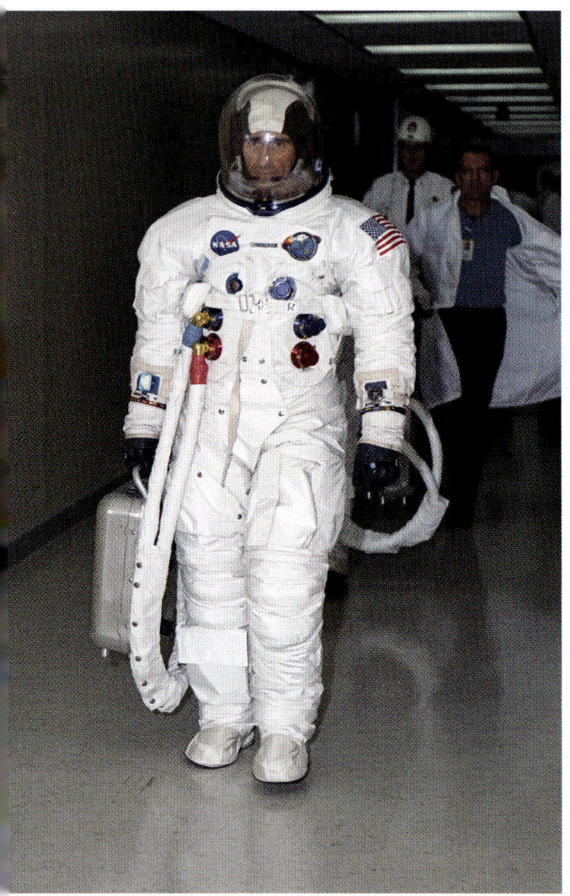

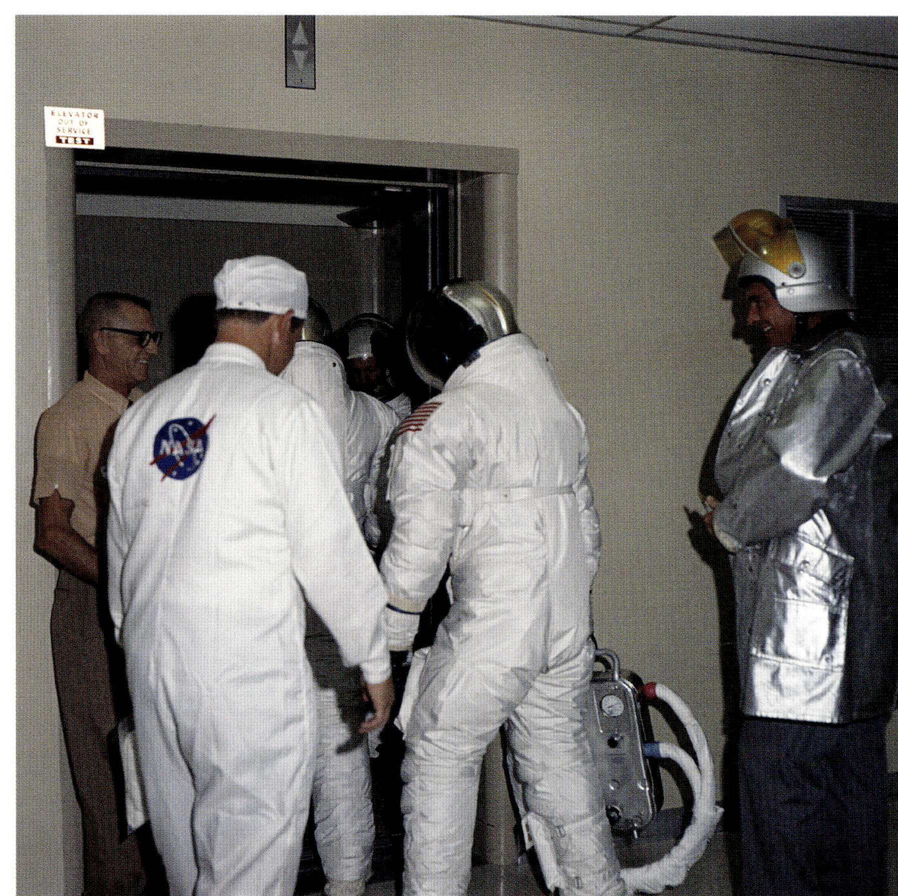

Cunningham heads for the elevator with Flight Crew Operations director Deke Slayton following.

The astronauts board the elevator. At left is Tony Broadway, MSOB supervisor. Firefighter Lee Ball, at right, would also serve as driver for an armored escape vehicle at pad 34.

Eisele leads Cunningham and Schirra to the Astronaut Transfer Van.

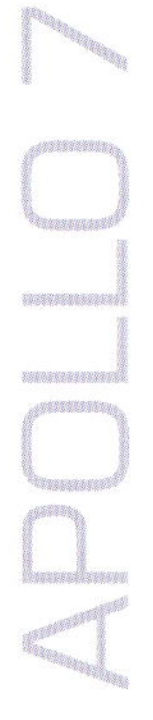

The van arrives at LC-34.

Schirra (*right*) is first out, followed by Teague and Cunningham.

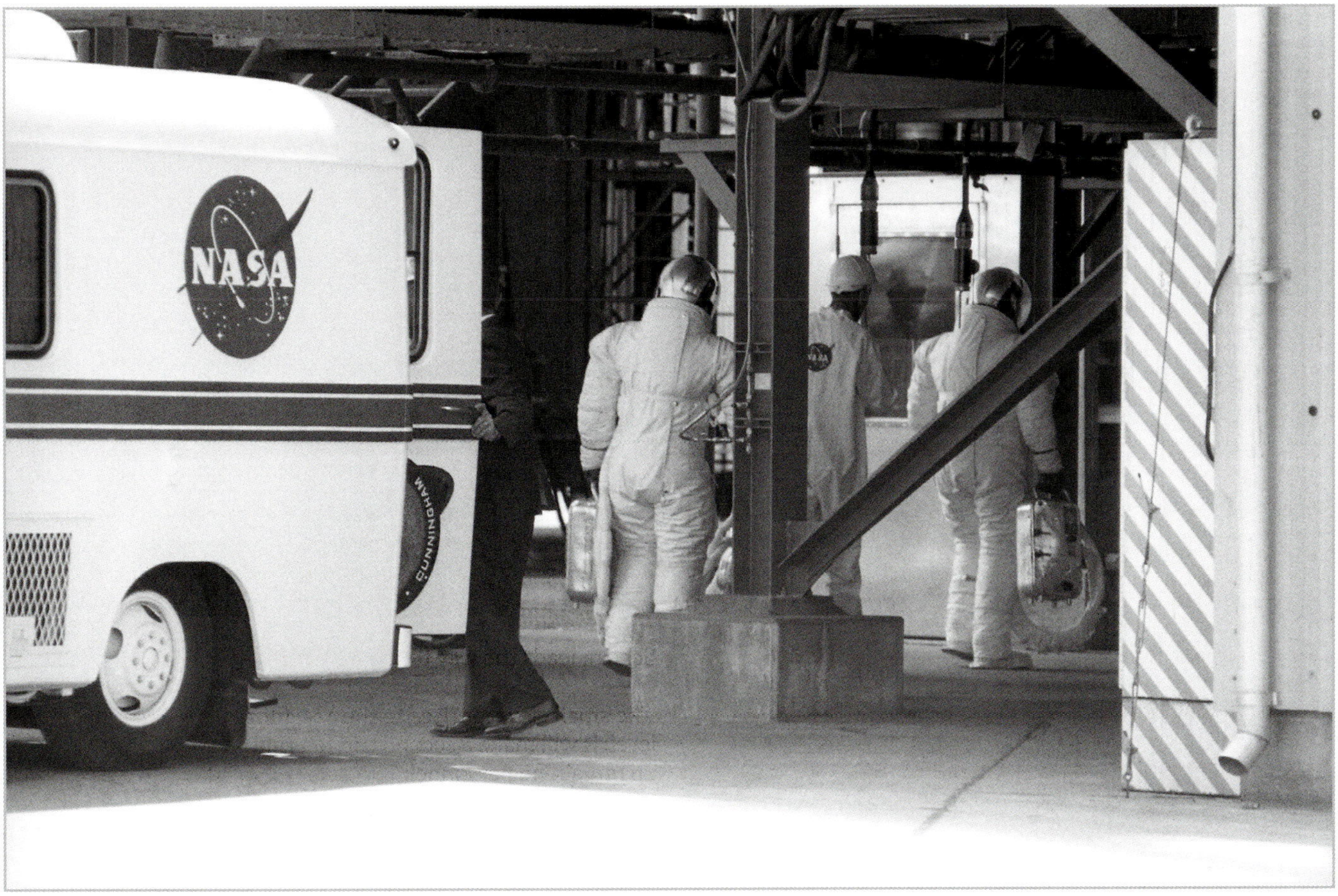

Schirra leads the way to the elevator.

The small elevator rises to the 224-foot level; it will make a second trip to bring up remaining personnel.

A camera captures the astronauts walking across the crew access swing arm.

Schirra (*left*) stops to look down.

Schirra (*left*) and Cunningham work with white room personnel.

Eisele (*left*) and Cunningham prepare for ingress.

The unaided emergency-egress drill is underway; Eisele (*left*) is the first to exit, with Schirra next.

Schirra and his crew walk under water sprays at the base of the pad toward rescuers. The three egressed from the CM and reached the ground in two minutes and five seconds, including a thirty-six-second elevator ride. It could have been slightly faster, Schirra tells reporters, but the astronauts were careful not to damage their suits.

An M113 armored personnel carrier takes the astronauts away from the pad.

The crew next takes the transfer van to the slide-wire termination area. Schirra is first out as NASA senior photographer Bill Taub (*right*) captures the scene. LC-37 is to the north (*at left*).

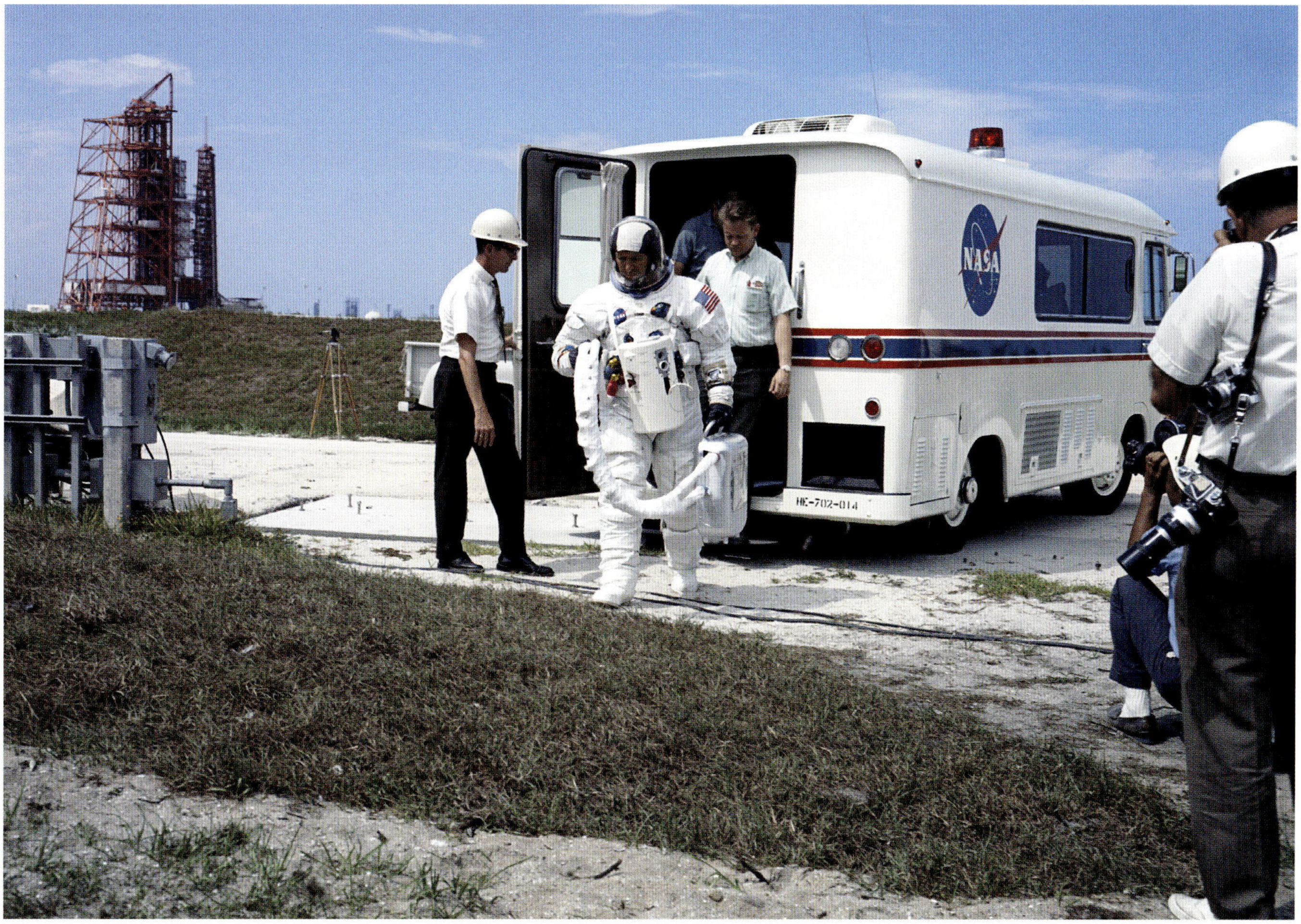

Schirra has an emergency air pack around his neck as a Bendix technician adjusts his sling.

Cunningham prepares to don a sling. The astronauts are shown how to detach the sling at the end of the trip.

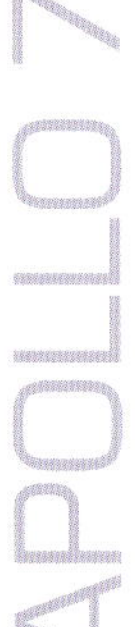

Cunningham dangles from the cable as Schirra (*left*) watches.

Eisele is trained on the sling. He will be first to come down the slide-wire.

A Hi-Ranger aerial crane next takes each astronaut to the top of the 35-foot training tower.

Walton works with Eisele's oxygen supplies. "This was our only chance to try it," he tells reporters. "At least I hope it is."

Eisele holds his emergency oxygen supply after his slide-wire run, with transfer van driver Steve Tatham at left.

Schirra is next to step off the tower.

Schirra zooms down the slide-wire, with Cunningham waiting in the crane at right.

Bendix personnel pull Schirra the final few yards to the end.

Schirra is back on the ground.

Schirra

Cunningham is the last of the three to slide down.

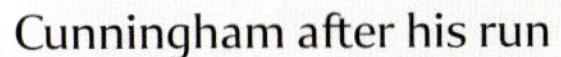

Cunningham after his run

Left to right: Cunningham, Eisele, and Schirra crouch for photographers at the beginning of their session with the news media. "Get a picture for the [NFL's Houston] Oilers," Schirra tells photographers.

The crewmen assume a three-point stance; they'd sign a copy of the photo to Oilers owner Bud Adams with the inscription "from eight short of eleven—Apollo Seven."

Left to right: Eisele, Schirra, and Cunningham coach the photographers, and Schirra scolds one for putting a lens on the sandy ground. “I wouldn’t do that with mine,” he says.

Eisele, Schirra, and Cunningham pose with the training tower, the crane, and the pad service structure behind them.

Eisele and Schirra. "The egress tests went very well today," Schirra says.

CBS News correspondent Steve Rowan (*right*) interviews the crew. Rowan would anchor the daytime CBS coverage from New York of the live Apollo 7 TV broadcasts.

Cunningham reflects on a question from Rowan.

The crewmen wrap up the session in good spirits. “The whole system is real good,” Schirra says.

Schirra and Cunningham answer questions from reporters. Schirra says the 35-foot ride from the training tower "was disappointingly short."

Schirra and Cunningham have their ventilators reconnected before reboarding the Astronaut Transfer Van.

Eisele and Cunningham enter the van.

Before the crew returns to the MSOB, Schirra talks with NASA engineering manager Jim Ragusa, who'd been first to test the full slide-wire less than a month earlier.

Pete Shores helps backup commander Tom Stafford with his life vest that afternoon in the MSOB as his crew gets ready for emergency white room egress training.

Stafford's suit is pressurized.

Stafford waits in a recliner.

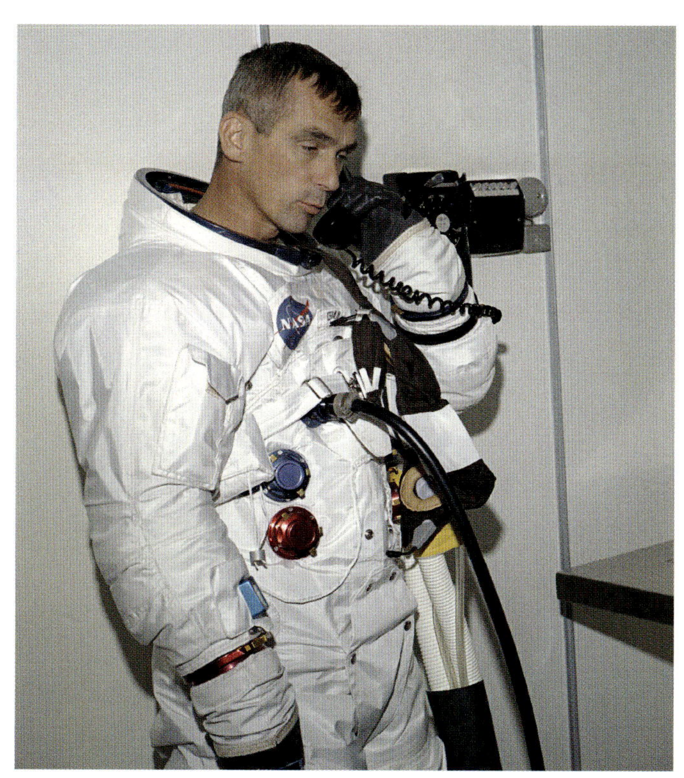

Gene Cernan on the phone.

Cernan bends as his suit is pressurized.

Cernan waits in a recliner.

John Young tests his mobility.

Cernan and Young arm wrestle.

Young, followed by suit tech Jim Lewis, leads Cernan and Stafford toward the transfer van. At LC-34, they perform only the unaided egress drill from the white room, not the slide-wire tests. Concerned about astronauts' safety during walkouts, NASA would install a metal ramp with hand railings over the entrance's two front steps in the next few weeks.

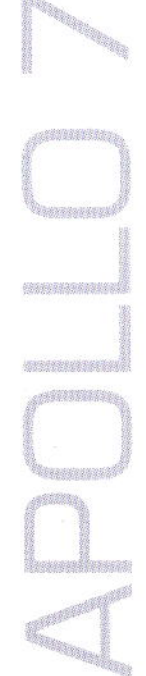

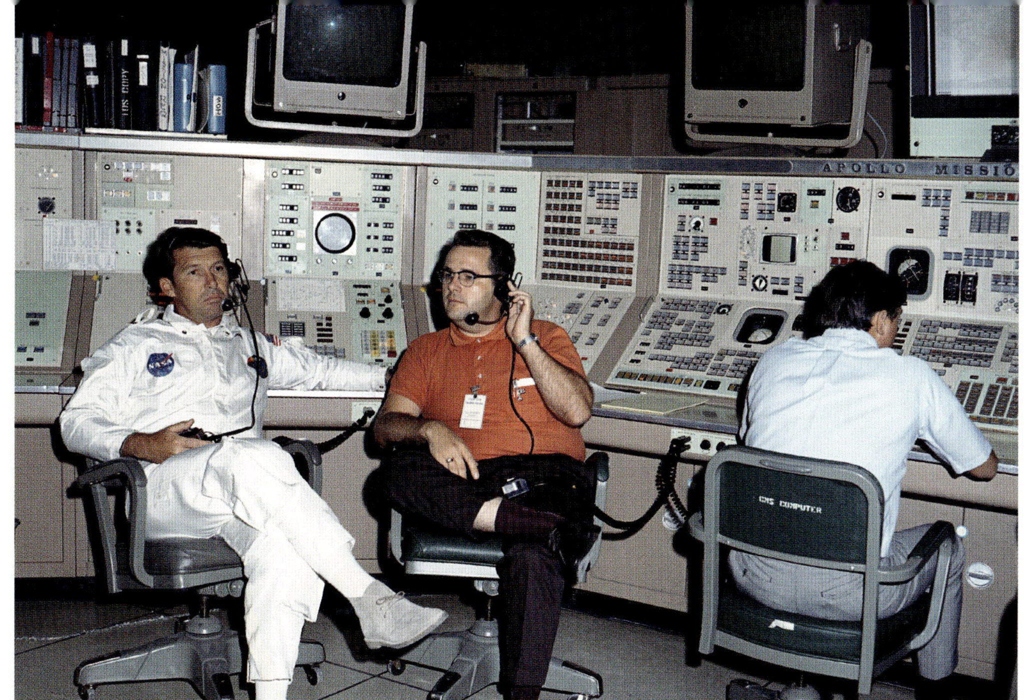

On September 12, Schirra (*left*) and Riley McCafferty, chief of the MSC Flight Crew Operations Branch, are part of a discussion at the Apollo Mission Simulator control station.

The prime crew had spent almost two hours in the simulator in the Flight Crew Training Building at KSC. Schirra wears a two-piece in-flight garment made of nonflammable Beta cloth.

The two share a laugh during the conference.

CHAPTER 7

September 15–October 10, 1968

Apollo 7 and its Saturn booster are illuminated by xenon spotlights at LC-34 late on September 15, during the first half of the Countdown Demonstration Test (CDDT). It had been halted the day before by two faulty spacecraft-fueling valves. This view looks northwest, with the umbilical tower behind the booster to the left. The top pair of service structure hurricane doors is retracted; each is 44 feet tall.

A second pair of hurricane doors protecting the S-IVB stage are retracted. The CDDT had begun on September 11.

The Countdown Demonstration Test (CDDT) was first implemented for AS-201, the first unmanned CSM test flight, in February 1966. It divided the practice countdown into two sections: "wet" (fueled) and "dry" (unfueled). The wet portion typically began five days before the targeted launch time and was carried out over two days, but without astronauts aboard. After the fuel tanks were drained, the dry part of the CDDT could begin. The crew boarded for the final few hours toward the mock liftoff time. The split-test concept was first developed during Gemini. This procedure continued through the space shuttle program, when it was renamed the Terminal Countdown Demonstration Test.

The eight extendable work platforms are being retracted from the spacecraft and booster before the service structure is rolled away. The white room and crew access arm are at center left.

A technician with Chrysler Corp.'s Space Division inspects the tail unit of the Saturn IB at fin no. 3. Each of the eight fins is more than 9 feet tall and 7 feet wide and provides aerodynamic stability during ascent.

The Apollo 7 CM remains unoccupied during this part of the test.

White room technicians secure the hatch on the CM.

The launch team monitors the countdown on the second floor of the LC-34 blockhouse, 1,000 feet from the pad. The first floor includes tracking, telemetry, and communications consoles. The interior dome was sprayed with a 2-inch coat of acoustical material, and the exterior is covered with reinforced concrete 5 feet thick, topped by a layer o f soil and more concrete.

Technician Roy Post in the blockhouse, officially known as the Launch Control Center.

A technician juggles phone calls. Flight controllers in the MOCR in Houston are online to support the test.

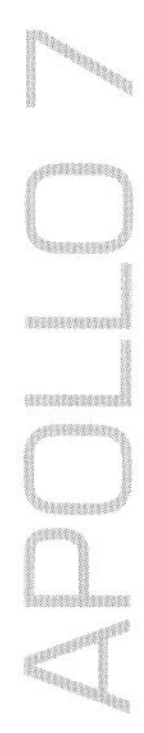

The work platforms continue to retract. The pad leader's desk is visible on Level 8 at center. The Q-ball at the tip of the LES remains covered.

With all hurricane doors and work platforms retracted, the service structure is ready to begin moving to its parked position.

A fish-eye lens captures blockhouse activity. The facility is designed to withstand a blast pressure equivalent to 50 kilotons of TNT at a distance of 50 feet.

KSC director Kurt Debus (*left*) and Apollo Spacecraft Program manager George Low confer.

KSC test planning manager Bob Moser had been test conductor for the launch of the first US satellite, Explorer I, and Alan Shepard's Mercury/Redstone-3 launch.

With work platforms retracted and all lines disconnected, the service structure begins to move southeast toward the camera on rails shortly after midnight EDT on September 16 in this view looking northwest.

CSM Systems Branch chief Arnold "Arnie" Aldrich stands near a periscope (*right*) with chief test supervisor Don Phillips behind him. Seated in front of the periscope is chief spacecraft engineer Ted Sasseen.

The high-pressure helium and nitrogen gas facility is illuminated as the service structure is rolled back. The gases are used for pressurization and purging of fuel lines. This view looks southeast.

A technician on the phone. The launch team wears Roanwell aviation headsets.

This view of the Apollo 7 Saturn IB looks northeast as the service structure rolls back early on September 16. At 2,800 tons and 310 feet tall, it is the largest movable land structure in North America.

Debus views the booster through a periscope. A second periscope and new consoles had been installed in 1963 to update the Saturn I facility for Saturn IB launches.

The Apollo 7 CSM and LES are exposed after the service structure has been rolled away, with the umbilical tower and white room at right.

This view looks northwest from inside the base of the service structure at its parked position 600 feet from the pad. It moves on four twelve-wheel "trucks" along a special dual-track railway within the complex. A 500 kVA diesel-electric generator supplies power to 100-horsepower traction motors in each truck.

KSC launch operations manager Paul Donnelly, a former Mercury chief spacecraft test conductor, monitors the test at his console.

KSC chief of public information Jack King talks with Debus. "The voice of launch control" during Gemini, King would provide launch commentary for Apollos 7–15, except Apollo 13.

A crescent Moon is above Apollo 7 as 41,000 gallons of RP-1 fuel and 66,000 gallons of liquid oxygen (LOX) fill its first-stage tanks. The second-stage tanks will hold 64,000 gallons of liquid hydrogen and 20,000 gallons of LOX. Fueling had begun at 5:00 a.m.

Left to right: Low, NASA launch operations director Rocco Petrone, and Donnelly. The team first wrestles with a short circuit in ground support equipment, soon followed by a faulty valve and a defective vacuum pump.

KSC deputy director Mike Ross (*left*) talks with Apollo program director Sam Phillips in the observation room of the blockhouse.

This view of the Apollo 7 Saturn IB from the service structure looks northwest.
Sunrise will be shortly after 7:00 a.m.

Preflight Operations Branch chief Ernie Reyes, Spacecraft Operations director John Williams, and Low monitor the progress of the CDDT.

NASA personnel at the booster consoles include Andrew Pickett (*pipe*), launch vehicle test operations manager. To his right is Hans Gruene, launch vehicle operations director. Ike Rigell, KSC launch vehicle operations chief engineer, is at far right.

The fueling test stretches into the afternoon, looking northwest from inside the base of the service structure.

The Apollo 7 CSM, SLA, and LES face the afternoon sun.

Test conductor Charles Stevenson checks on the launch vehicle with a periscope.

IBM engineers deal with a final glitch just ten minutes before the planned "launch." Computers in the IU and the blockhouse fail to communicate, but the problem is quickly resolved and the count resumes at 6:04 p.m.

Aldrich (*right*) speaks during the count.

Kurt Debus, George Low, and Rocco Petrone are pleased after the "launch" at 6:15 p.m., although it comes seven hours behind schedule and more than two hours past the planned T-zero. That same day, NASA administrator Jim Webb announces his retirement effective October 6, citing budget reductions.

Deputy launch director Walt Kapryan talks with Low after the test.

Chief test supervisor Don Phillips (*left*) with Low.

The next day, the Astronaut Transfer Van waits outside the MSOB for the Apollo 7 astronauts who will take part in the dry (unfueled) portion of the CDDT, which had begun at 8:00 a.m.

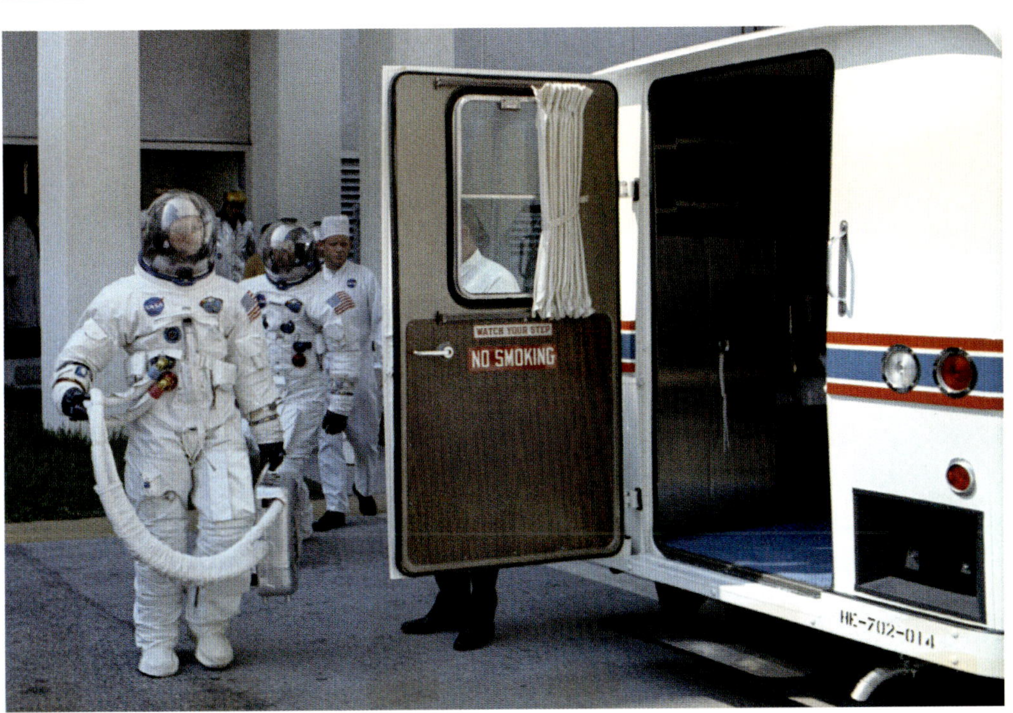

Eisele leads Cunningham toward the van.

Eisele prepares to enter the van, with Cunningham behind him.

Eisele enters the van, followed by Cunningham, with Schirra bringing up the rear.

Schirra is the last crewman in, followed by suit tech Clyde Teague. Driver Steve Tatham holds the door.

Deke Slayton and KSC security chief Charles Buckley prepare to board, with KSC firefighter Joe Morgan at right. The astronauts ingress the CM just before noon, and the mock launch comes at 3:20 p.m.

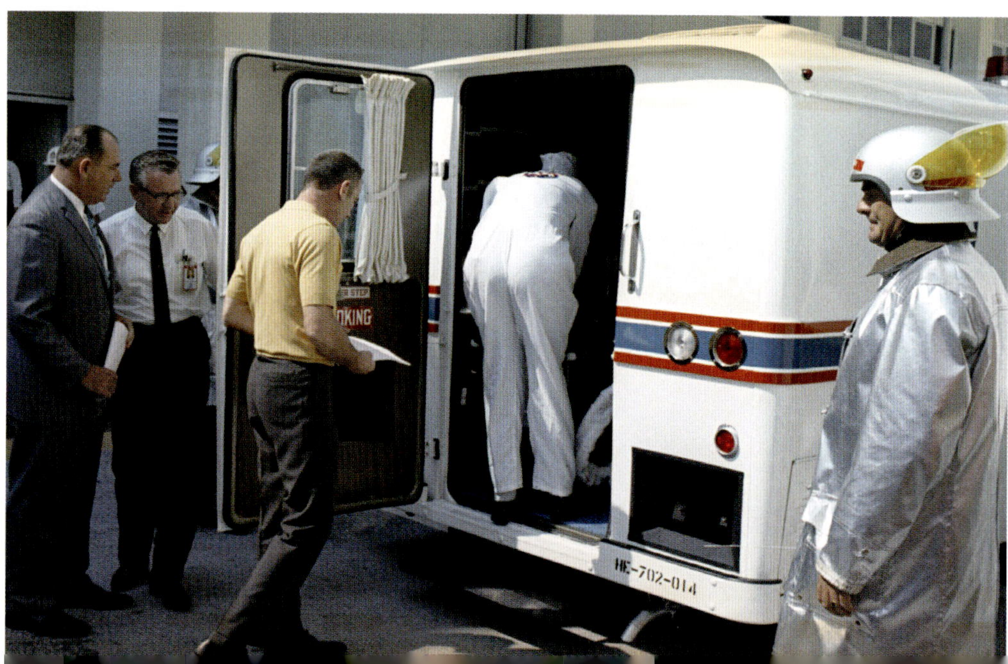

Left to right: Schirra, Eisele, and Cunningham answer questions during their preflight news conference in the Building 2 auditorium at MSC on September 20. "We see no reason for us to delay beyond the announced [launch] date," Schirra tells reporters. "We are just about ready to fly the mission." The Saturn IB model displayed is in the SA-6/SA-7 configuration.

"I feel it is a very ambitious undertaking," Schirra adds. "We have had a goal that has been a particularly hard one to achieve, particularly when we have to follow one where we lost our compatriots." His disclosure to reporters that afternoon that he planned to retire as an active astronaut after the mission, however, is what makes headlines.

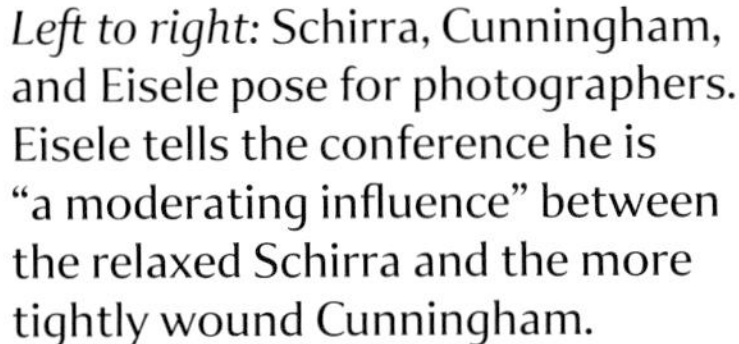

Left to right: Schirra, Cunningham, and Eisele pose for photographers. Eisele tells the conference he is "a moderating influence" between the relaxed Schirra and the more tightly wound Cunningham.

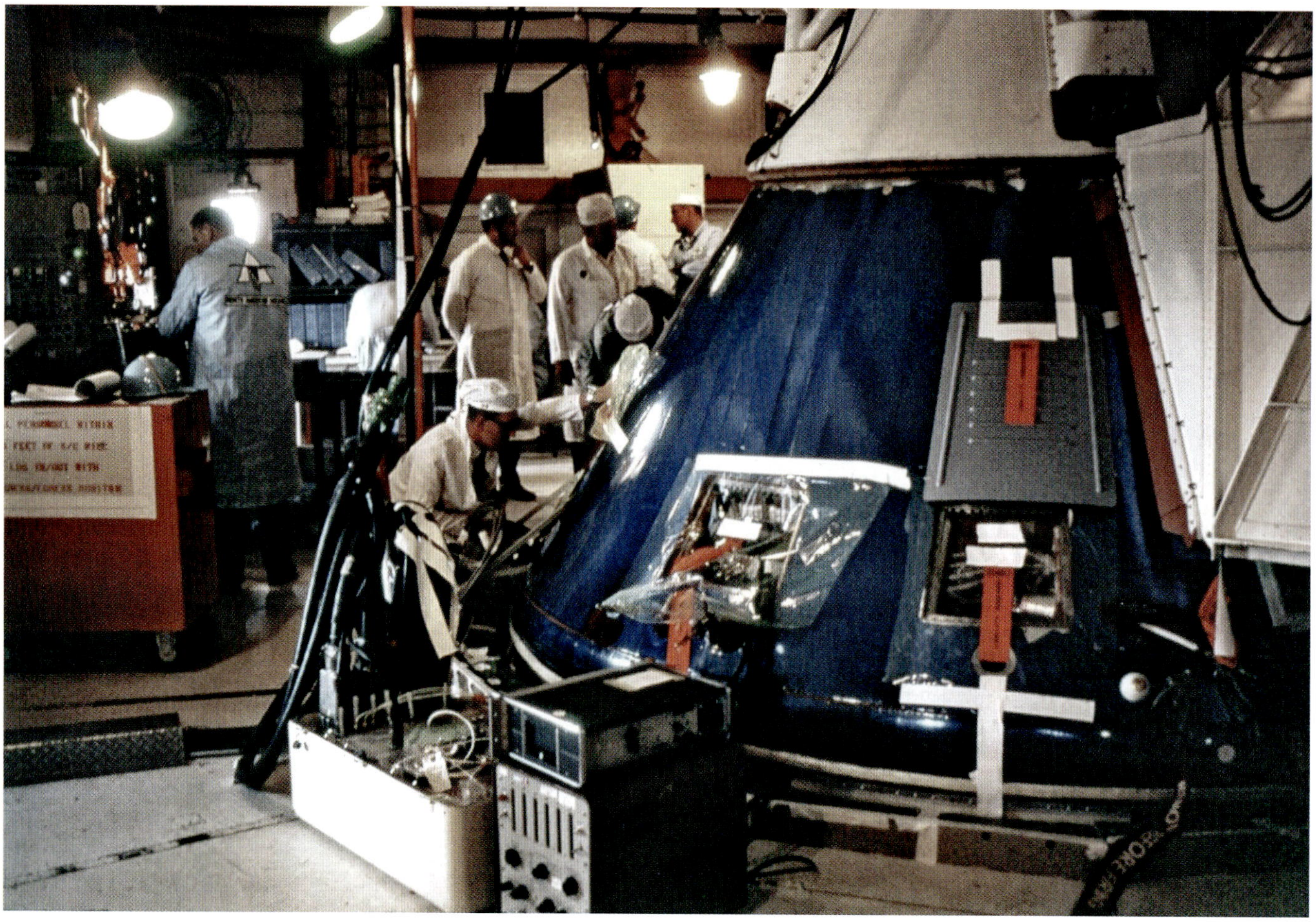

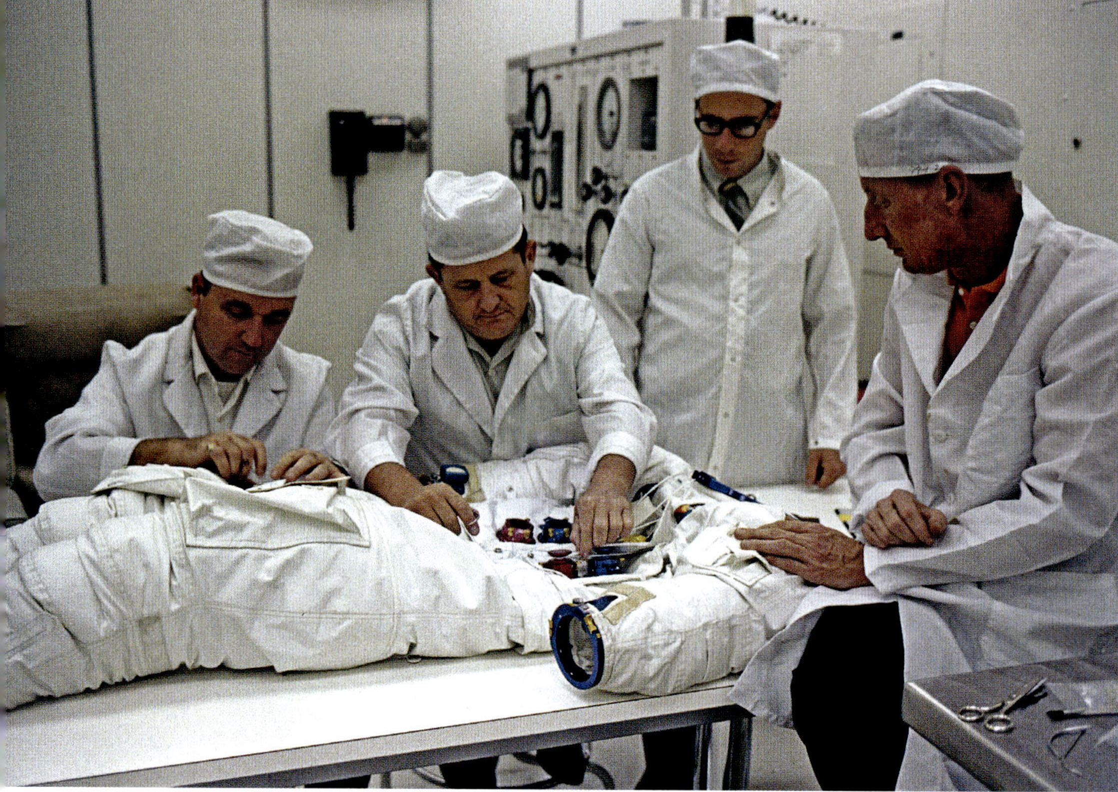

RCS lead engineer John Tribe checks on a CM RCS access panel on Level 8 at the pad in late September. A portion of the Boost Protective Cover has been removed; the white room is in place at right.

Suit techs make final pressure checks in the MSOB on October 8. *Foreground, left to right*: Bill Dunsworth, Andy Wolf, and Clyde Teague.

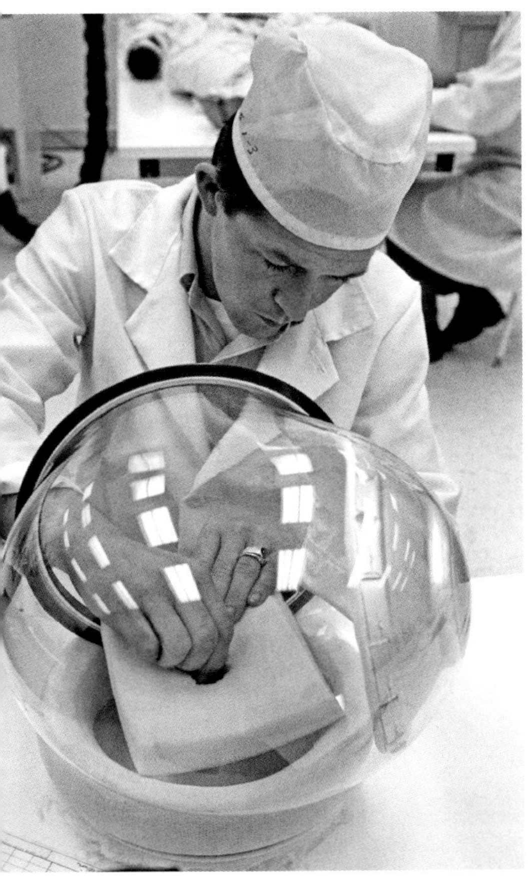

Suit tech Bill Prior cleans the inside of one crewman's helmet on October 8.

Left to right: Schirra, Cunningham, and Eisele chat near the AMS in the Flight Crew Training Building on October 8.

Cunningham (*center*) makes a point to Eisele as Schirra listens.

Eisele talks with AMS engineer Robert Kain.

Left to right: Eisele, Schirra, and Cunningham discuss a weather map with Slayton in the conference room of the crew quarters in the MSOB on the morning of October 9. They also get in some simulator time before Schirra goes dove hunting with friends in the afternoon, while Cunningham and Eisele watch the World Series on TV. Also, that morning, Apollo 8 is rolled out to LC-39 from the VAB.

Eisele with a flight document. A countdown clock is on the wall (*at upper right*).

The Apollo program director, Lt. Gen. Sam Phillips, reads from a document, which draws laughter during a meeting with the astronauts on October 10, the day before launch. *Clockwise from top*: Schirra, public affairs officer Paul Haney, CSM manager Ken Kleinknecht, Phillips, Alan Shepard (*behind Phillips*), Low, Eisele, MSC director Robert Gilruth, unidentified, Petrone, MSC director of medical operations Dr. Charles Berry, and Cunningham.

Schirra welcomes MSFC director Wernher von Braun; Debus is behind him. Slayton (*seated, center*) has also arrived.

Dr. Berry, Cunningham, and Schirra share a laugh.

Shepard shows Schirra a photo of the Apollo 7 / Saturn IB at LC-34 from the September 16 CDDT.

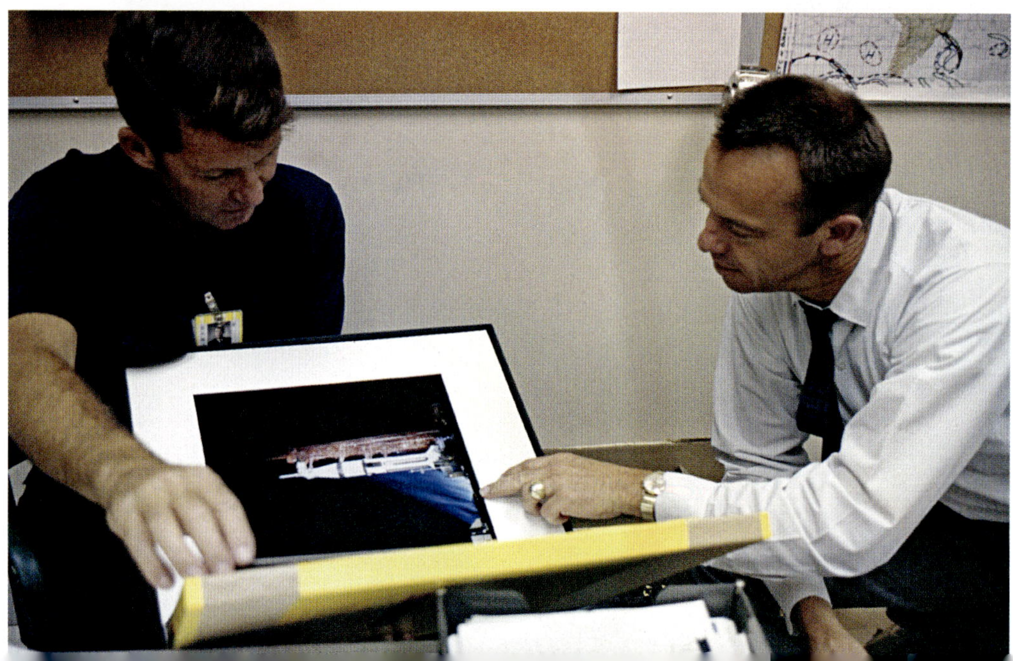

Tourists at the KSC Visitor Information Center inspect a CM mockup in October 1968 as interest builds for the first manned Apollo mission.

Correspondent Walter Cronkite, holding a CSM model made for NAR by the Walter J. Hyatt Co., will anchor TV coverage of Apollo 7 for CBS News. *CBS photo*

NBC News anchors Chet Huntley (*left*) and David Brinkley will be joined by Rene Carpenter, wife of former astronaut Scott Carpenter, as a special correspondent intended to attract more women viewers. *NBC photo*

Science editor Jules Bergman and evening news anchor Frank Reynolds will cover Apollo 7 for ABC News. *ABC photo*

Bergman previews the launch with two Hyatt CSM models on the roof of the ABC News building at the press site just north of the Cape Kennedy AFS Industrial Area, which had been used for Gemini launches. The tempered-glass panel behind him is a light diffuser. *ABC photo*

Former NAA test pilot George Smith serves as a crewman in a full-scale CM mockup as part of ABC News launch coverage at the Cape. *ABC photo*

CHAPTER 8

October 10–11, 1968

Two pairs of hurricane doors are retracted, exposing the CSM and the S-IVB in the LC-34 service structure on the evening of October 10, with launch set for the next morning.

Work platforms are retracted; the pad leader's desk on Level 8, to the left of the LES, has been cleared.

Chrysler technicians work at the tail section of the S-IB stage. *At left*, a technician looks down from an access door in the tail shroud. Two of the eight hold-down arms are at center and far right, with one of two short cable masts between them. The masts, which carry electrical cables and pneumatic service lines, retract when the booster is 4 inches off the pad.

The service structure begins to roll away just before 10:00 p.m. EDT. Backup CM pilot John Young enters the CM to check switch positions.

The structure moves toward its parked position in this view looking west.

An amber light (*at left*) signals caution. Launch preparations continue until 11:00 p.m., when a built-in six-hour hold begins, and the pad is cleared. The pad's main LOX tank, 43 feet in diameter, is dimly illuminated at right.

Two of the xenon spotlights providing illumination are in the foreground. The trouble-free countdown resumes at 5:00 a.m. EDT on October 11, when tanking begins.

The Apollo 7 / Saturn IB space vehicle vents LOX, with the blockhouse in the foreground, as fueling continues.

KSC director Kurt Debus sits between deputy launch director Walt Kapryan (*rear*) and launch director Rocco Petrone (*front*) in the blockhouse.

Breakfast guests in the crew quarters dining room include just-retired NASA administrator Jim Webb (*left*) and KSC mission support chief Hal Collins (*standing behind Slayton at right*). NASA's NAR representative Fred Peters is at the end of the table at left, sitting across from CSM manager Ken Kleinknecht. Crew quarters cook Lew Hartzell, behind Cunningham, brings more coffee. Technicolor's Larry Summers films.

Cunningham

Schirra (*left*) and Slayton enjoy New York strip steaks, eggs, toast, coffee, and orange juice. Slayton had awakened the crew at 6:00 a.m.

Kleinknecht and Eisele

Eisele reads *Today*, the local newspaper. Cunningham later called the big headline "glorious." Eisele's last name is often mispronounced, leading to his gag coffee mug.

Left to right: Kleinknecht, Eisele, Schirra, and Slayton, and support crew members Pogue and Evans, who participated in the emergency slide-wire tests

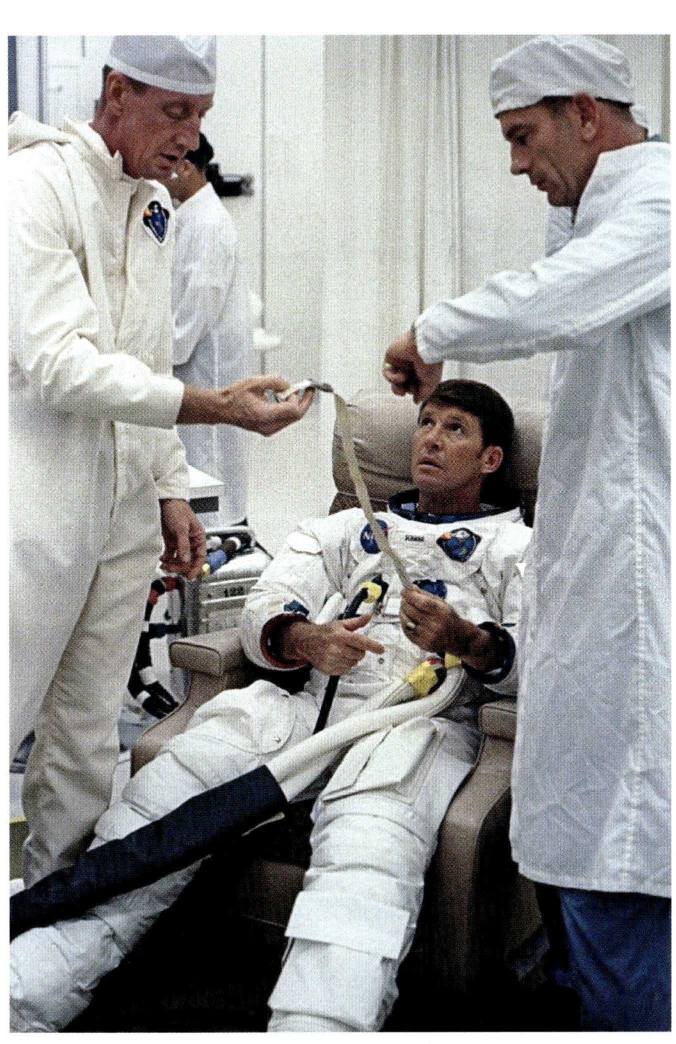

Suit tech Clyde Teague (*left*) helps Schirra in the suit room on the third floor of the MSOB, with Slayton at right. The flight will be the first test of new lunar-type A7L pressure suits manufactured by the International Latex Corp. (ILC).

The International Latex Corporation (ILC) of Dover, Delaware, provides the Apollo pressure suits after winning a competition with Hamilton Standard and the David Clark Co. Each crewman has three custom suits: one for flight, one for training, and one for backup. The suits are fabricated from fourteen layers of aluminized Kapton, neoprene-coated nylon, an outer layer of flameproof Beta cloth, and wear-resistant Chromel-R fabric at the joints. The astronauts wear a "long johns"-like liquid-cooled undergarment. The helmet has a transparent polycarbonate shell with a feed port at the front and a vent pad at the back, through which oxygen flows to the face area. Each 70-pound suit costs about $100,000.

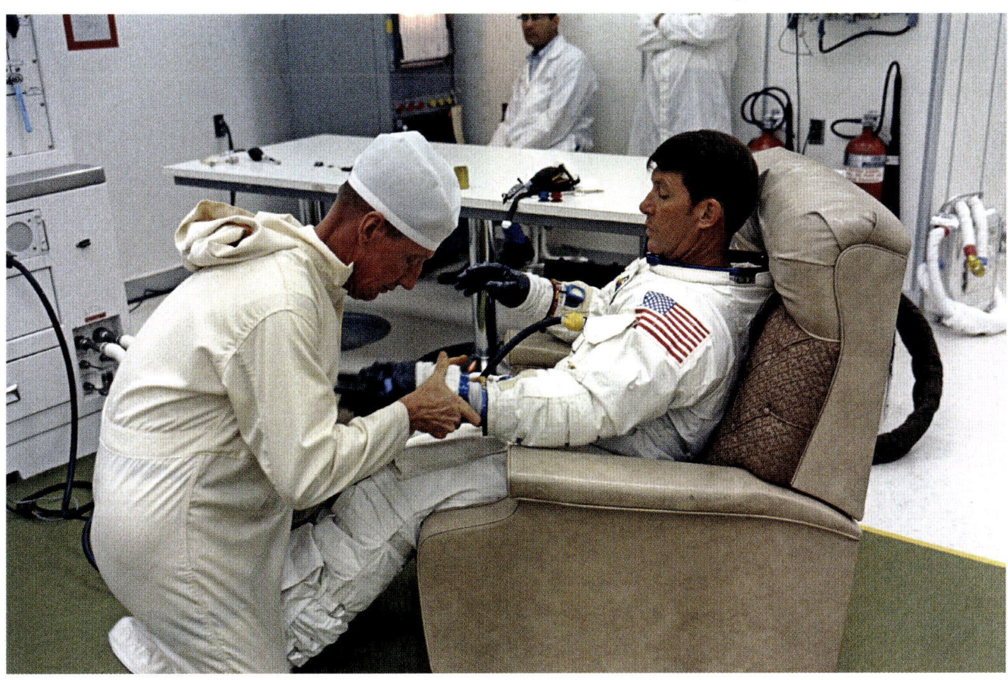

Teague attaches Schirra's left suit glove as technicians in the background monitor the astronaut's biomedical sensors.

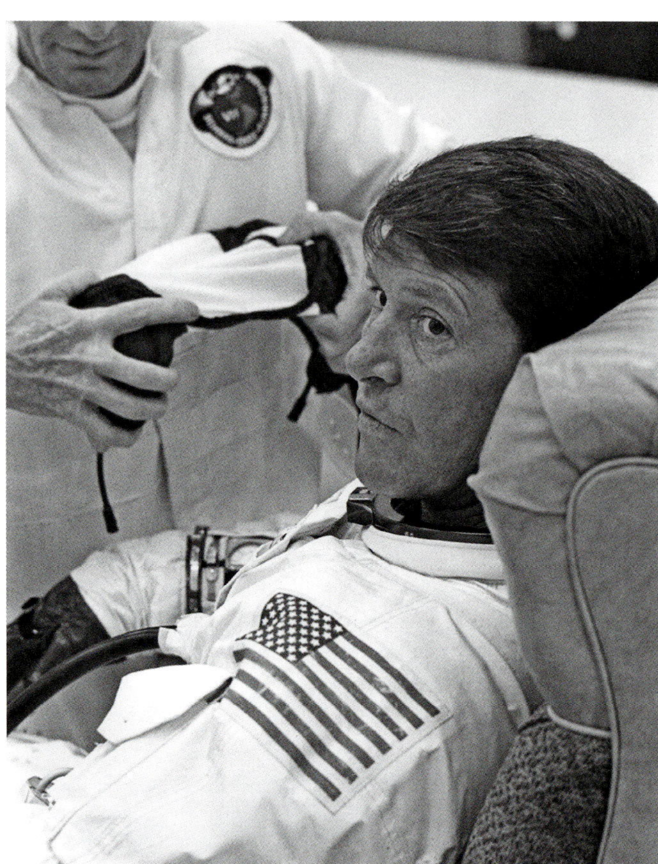

Teague readies Schirra's communications carrier (headset).

ILC suit tech Jim Lewis works with Cunningham's cuff. The crewman's inlet and outlet oxygen suit hoses are attached, as is the black electrical and communications cable.

The astronauts' communication carrier, with its twin mics and chinstrap, becomes nicknamed the "Snoopy cap," which resembles the white head and black droopy ears of the *Peanuts* comic strip dog Snoopy.

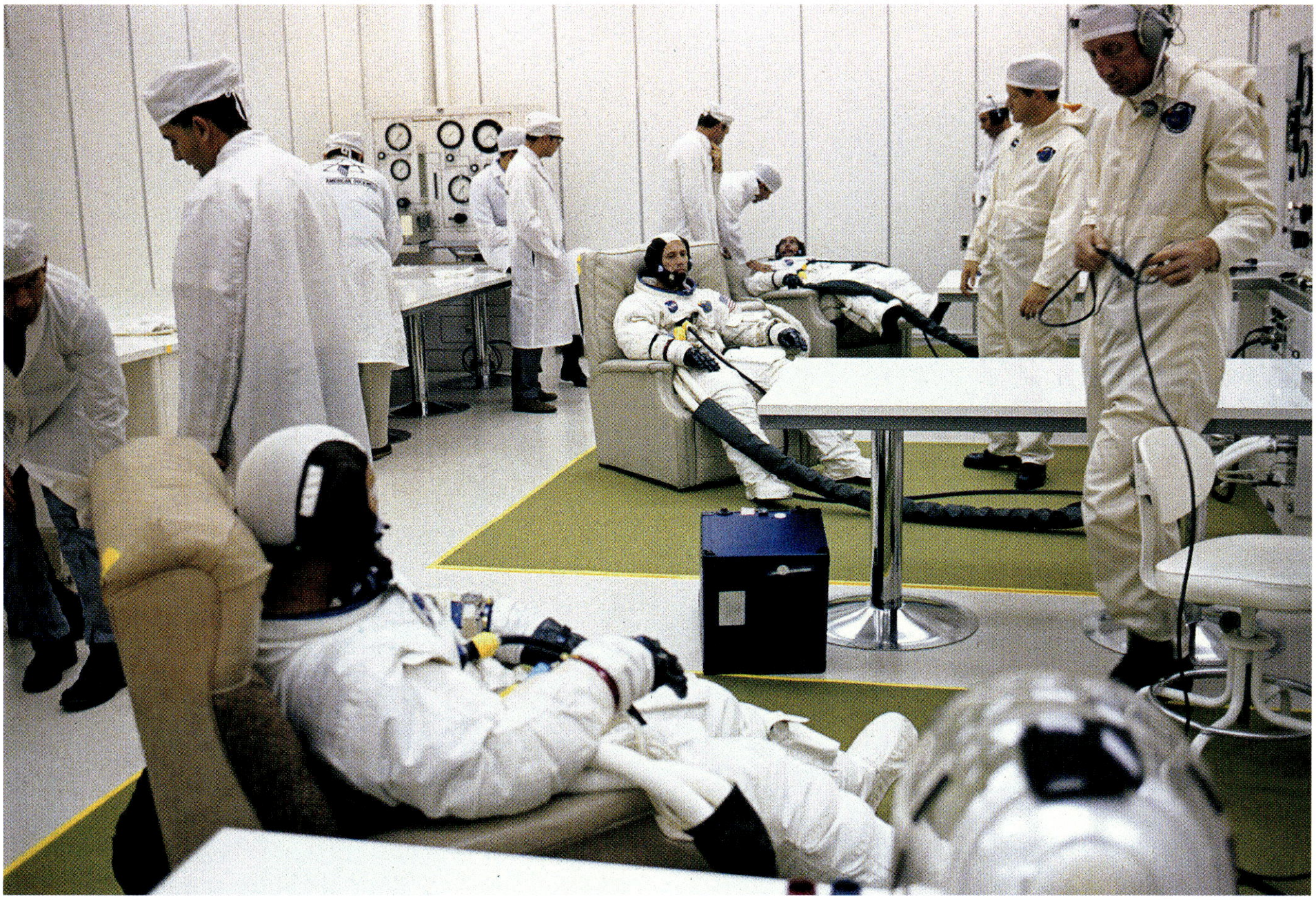

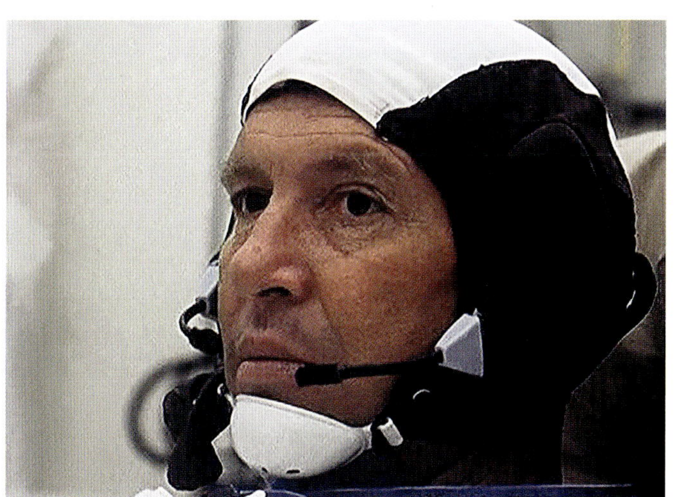

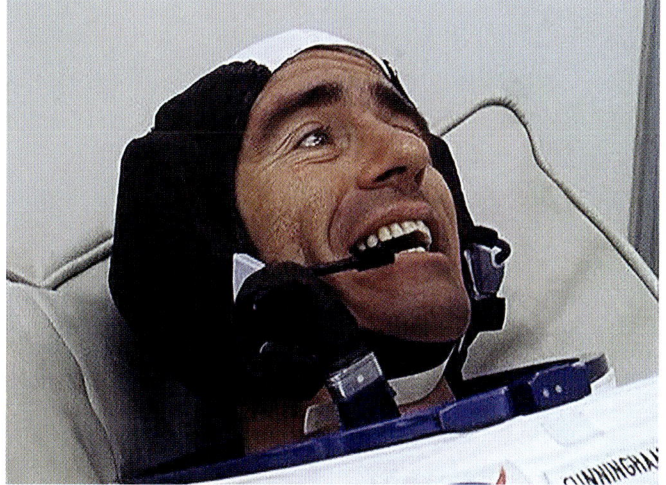

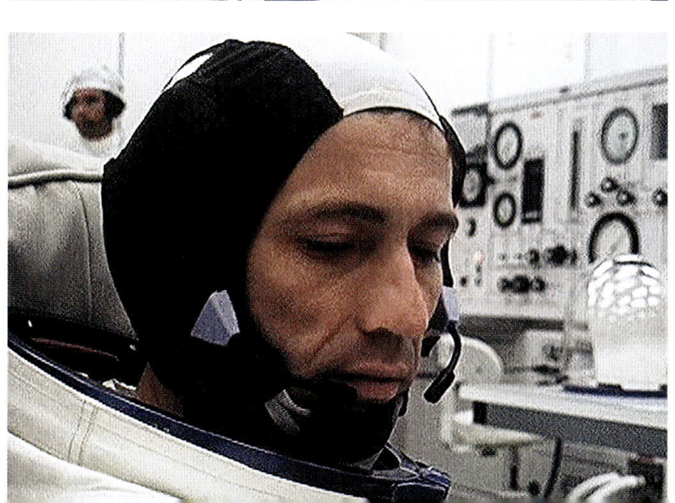

Suit room activity involves technicians from NASA, ILC, and NAR.

Schirra is reflective as launch approaches.

Cunningham (*right*)

Eisele (*left*)

Slayton talks with Eisele as Cunningham reclines (*at right*).

Cunningham

Suit tech Jim Lewis talks with Eisele about his gloves.

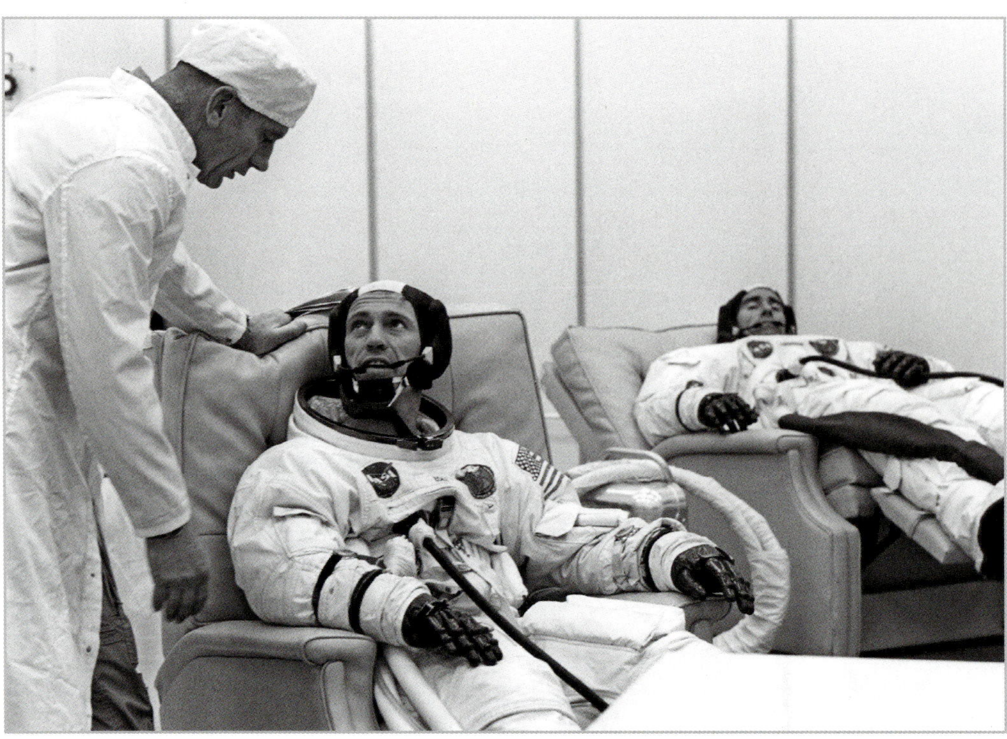

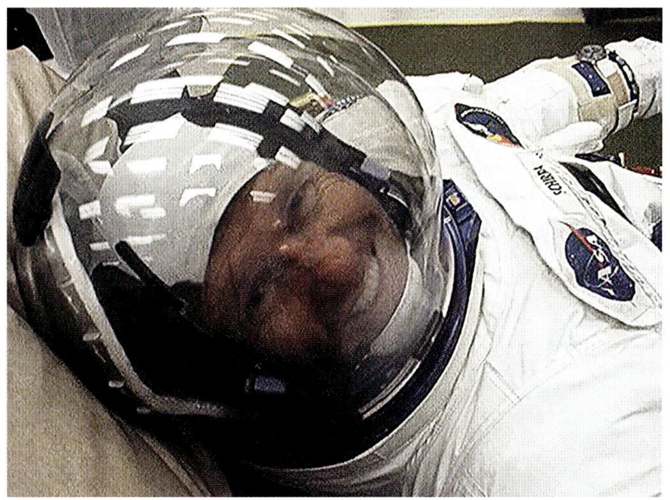

Teague (*out of frame*) adjusts Schirra's neck ring before he dons his helmet.

Schirra smiles through his helmet. The new bubble helmets are made of polycarbonate, thirty times stronger than the Plexiglas used for Mercury and Gemini visors.

Cunningham attaches his chinstrap.

Schirra's suit is pressurized. The crewmen also begin breathing pure oxygen to purge their bloodstreams of nitrogen, to prevent the bends as the cabin pressure drops during ascent.

Eisele and Cunningham have their suits pressurized and begin to prebreathe oxygen.

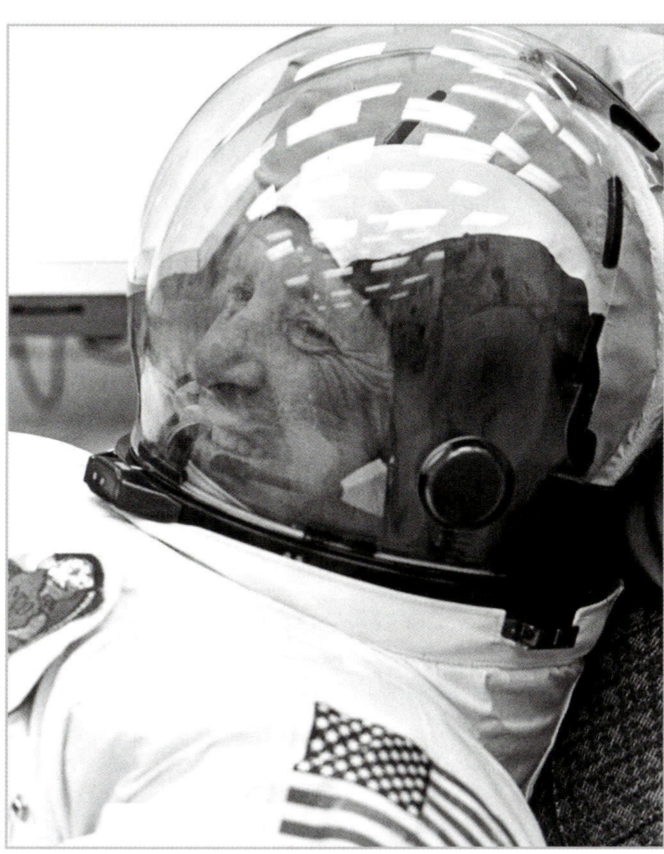

Schirra clowns menacingly behind Slayton as Cunningham and Teague follow in the MSOB hallway to the elevator.

Eisele approaches the Astronaut Transfer Van, followed by Walton, Cunningham, and Schirra. KSC security chief Charles Buckley holds the door.

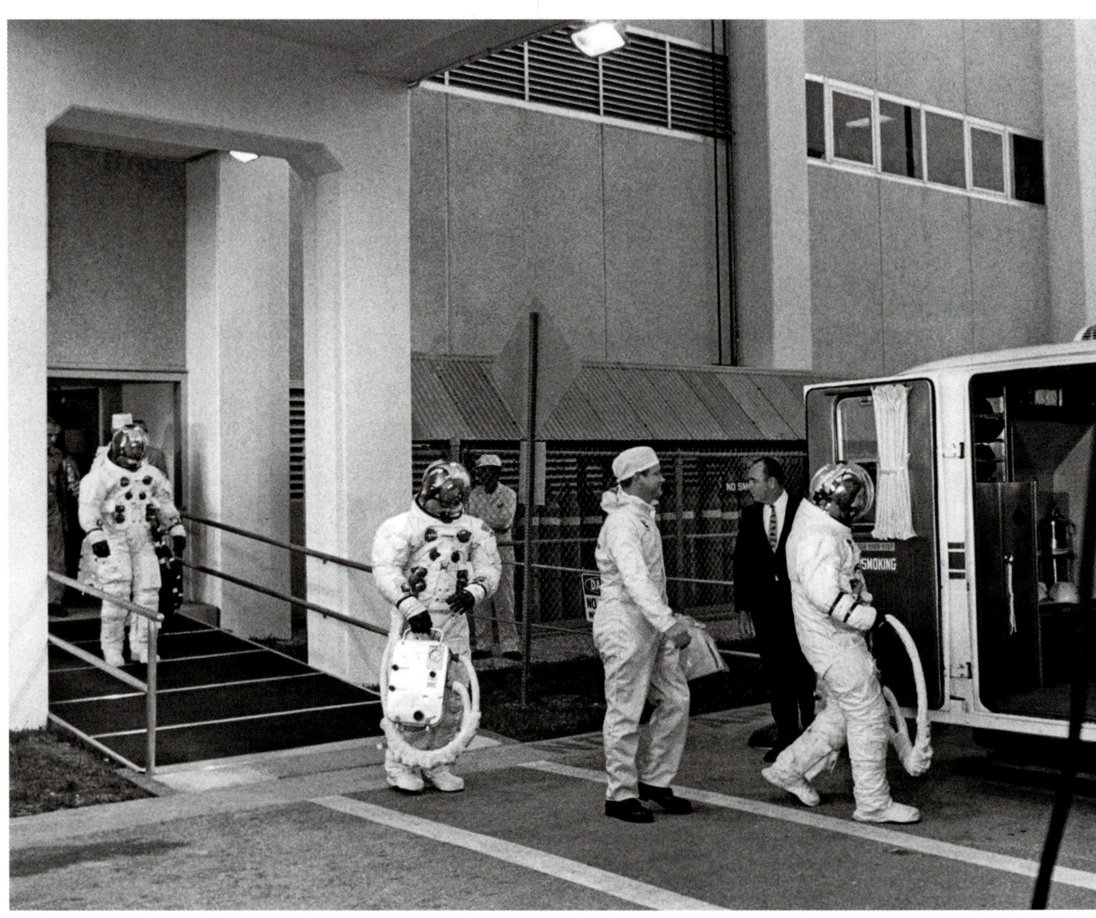

Eisele near the van. MSOB supervisor Tony Broadway watches in the background.

As the last to enter, Schirra waves to reporters and hundreds of space workers at 8:10 a.m. At the pad, with fueling complete, the closeout crew can return to the white room.

Schirra is first out at the pad about 8:20 a.m., in a photo from the still camera in the van's rear compartment.

Cunningham (*right*) follows Schirra.

Schirra (*right*) and Cunningham approach the umbilical tower elevator; they will ride up together with one suit tech, who will send the elevator back down for Eisele.

Eisele, who waited in the van, heads for the elevator as suit tech Walton, who'll join him, brings a backup ventilator.

Schirra waits to enter the CM first at 8:30 a.m. Backup CM pilot John Young, who'd earlier been checking CM switch positions, stays to help with crew ingress.

Cunningham will next ingress into his right-hand couch.

NAR pad leader Guenter Wendt talks to Eisele before he takes the center couch.

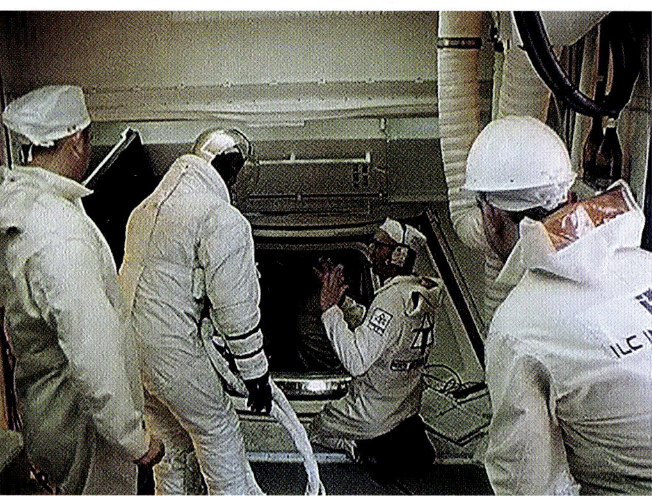

The team pushes firmly to close and latch the new crew compartment hatch at 9:15 a.m. Young uses the overhead grab bar.

Before closing the BPC hatch, the closeout crew begins to remove a portion of an overhead panel at 10:00 a.m., to prepare the white room to partially retract. Teague stands at left, Wendt at right.

As daylight floods the white room, the BPC hatch is closed and the technicians depart at 10:20 a.m. "Guenter went," says Eisele.

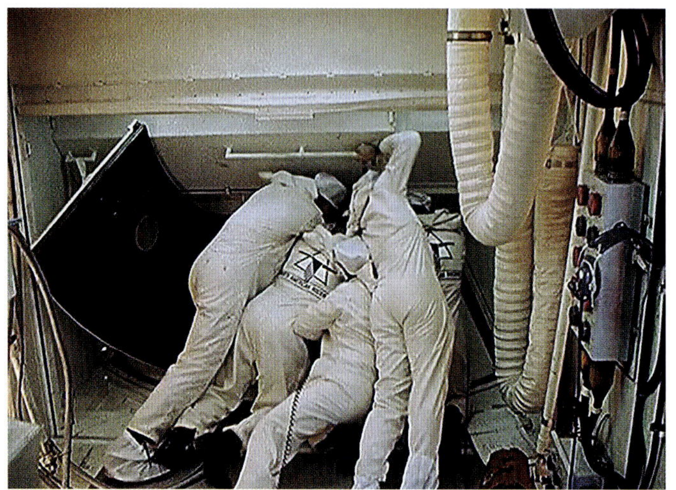

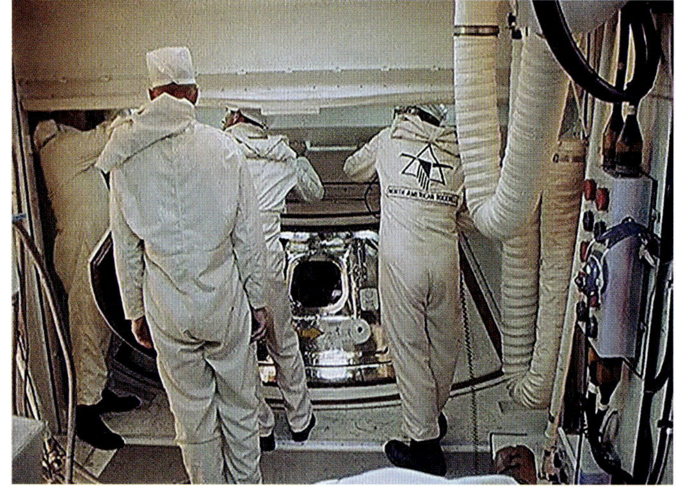

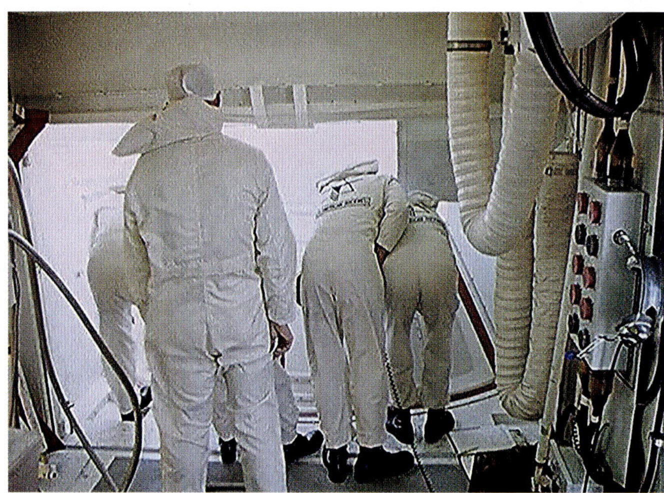

The Saturn's stages vent oxygen in this view looking northwest. The pad's spare flame deflector is parked between the service structure (*at left*) and the booster, with the LOX storage tank at right. "From what I can see, it's as blue as a bluebird out there," Schirra reports.

Flight directors Glynn Lunney (*left*) and Gerry Griffin at the MOCR at MSC in Houston monitor the countdown. A brief hold is called with six minutes and fifteen seconds to go, when the S-IVB's J-2 engine's thrust chamber hasn't cooled sufficiently, but the glitch is quickly resolved.

Journalists and technicians wait at the press site southwest of the pad, which had been used for Gemini launches. The ABC News area includes a full-scale CM mockup.

Buzz Aldrin, backup Apollo 8 CM pilot, talks with Webb at the VIP site on the East NASA Causeway along the Banana River. “This is the first time since [Gus] Grissom in the Redstone I haven’t been working,” he says.

Left to right: Susan Borman, Apollo 8 commander Frank Borman, and Jerry Carr at the VIP site. Carr is on the Apollo 8 support crew.

Apollo 8 LM pilot Bill Anders (*right*) and his backup Fred Haise are ready to photograph the launch from the VIP site.

US representative Olin Teague (D-Texas), chairman of the House Committee on Science and Astronautics, tells Webb he's "sorely disappointed by budget cutbacks in the space program," which Webb had cited for his resignation.

Spectators wait at the viewing stand. More than a thousand launch guests watch from the VIP site, but Cunningham's wife, Lo Ella, and their children, Brian, eight, and Kimberly, five, are on a boat in a river with a TV. Jo Schirra and Harriet Eisele stay in Houston.

Florida governor Claude Kirk (R) is interviewed by an AP reporter at the press site. The governor is also interviewed live by Walter Cronkite in the CBS News trailer. Kirk was disappointed that his 1966 campaign prevented him from seeing the Gemini XII launch, but he did attend the second Saturn V test flight in April.

At the VIP site, comedian and actor Bill Dana, with Kirk, holds a commemorative postal cover and a 16 mm camera. Dana's most famous character is José Jiménez, "the reluctant astronaut," and he had become friends with Schirra and the other Mercury astronauts.

Dana films Kirk, with Willard Rockwell, NAR chairman and president, to his right. Dana's record producer, Michael Knapp, provides the title cards the crew will use during their TV downlinks.

Launch guests include Apollo I astronaut Ed White's brother and father, the prime minister of New Zealand, World War I ace fighter pilot Eddie Rickenbacker, former "voice of Mercury Control" John "Shorty" Powers, and Gen. Curtis LeMay, who was then the vice presidential running mate of George Wallace on the American Independent party ticket. "I'm not here to answer questions," he snaps.

Backup crewman Gene Cernan and his wife, Barbara, train cameras on the Saturn IB. The astronaut, who'd been in the AMS until midnight, welcomes the break. "This is the fun part of our work," he says.

An ice coating falls from the supercold S-IB LOX tanks at the 11:03 a.m. EDT ignition. The white room is at the end of the crew access arm (*at upper right*); it had been fully retracted five minutes before launch.

Bendix workers along Cape Road southwest of the pad watch the liftoff. Gusty winds had diminished in time for liftoff.

The Saturn IB is framed at ignition by the service structure parked 600 feet behind it.

The 1.3-million-pound Apollo/Saturn clears the umbilical toward within seven seconds, and control switches to the MOCR at MSC.

UNITED ST
UNITED STAT

ED STATES
ED STATES

Apollo 7 launches from the same pad used for Apollo I eighteen months earlier. It is also the final of seven launches from LC-34, and the first time that astronauts ride a member of the Saturn rocket family.

The RP-1 kerosene fuel used in the first stage produces a yellowish exhaust.

This view looks north from LC-13 as Apollo 7 launches. Apollo 8 and its Saturn V, rolled out two days earlier, are in the distance at left at LC-39A. LC-37 and its service structure and twin umbilical towers are in the middle. LC-14 is in the foreground.

At T-minus three seconds, the first stage's eight Rocketdyne H-1 engines ignite in pairs, 100 milliseconds apart.

Apollo 7 is the final manned launch from Cape Kennedy until 2024, since all future crewed missions through the space shuttle program will use LC-39 at KSC on adjacent Merritt Island.

A launch photo triptych, looking north

At the White House, President Lyndon Johnson watches the launch on three color TVs in the Oval Office with the foreign minister of France, Michel Debre (*not seen*).

Apollo 7 clears the umbilical tower, with the Atlantic Ocean beyond. It is the most powerful rocket in the world, developing 1.6 million pounds of thrust at launch.

A camera aboard a USAF Boeing C-135B Stratolifter flying at more than 35,000 feet, based at Patrick AFB, captures the launch with the VAB, using an Airborne Lightweight Optical Tracking System.

A tracking camera follows the Saturn IB during ascent. "She's running—it's getting a little noisy now," Schirra radios.

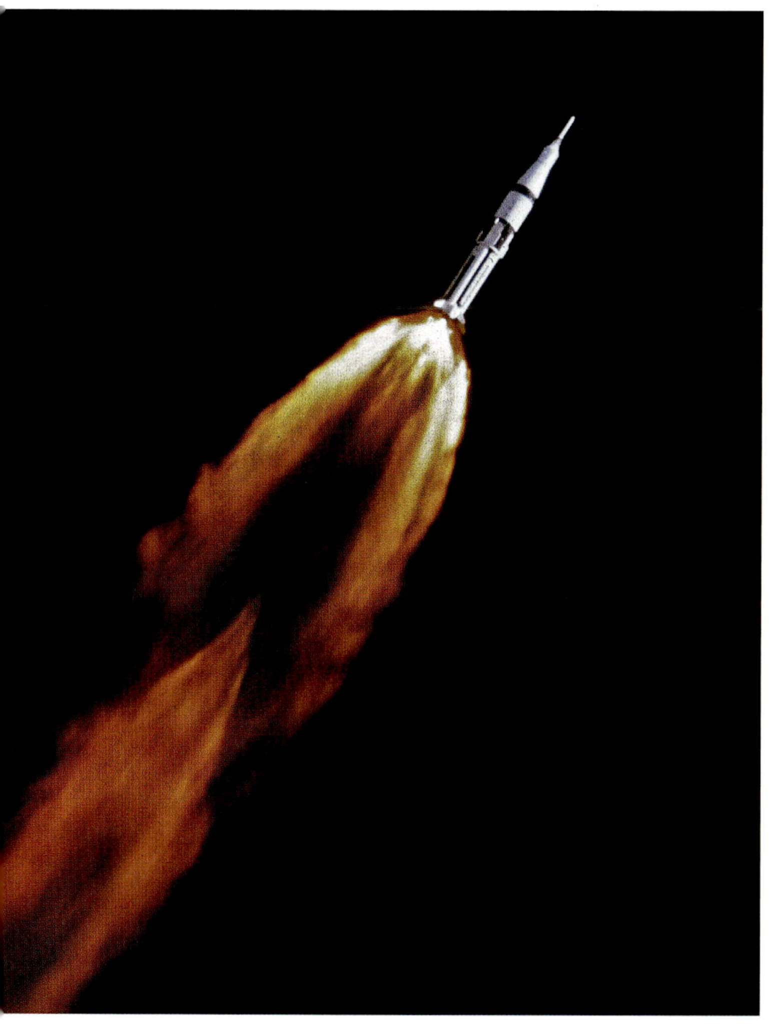

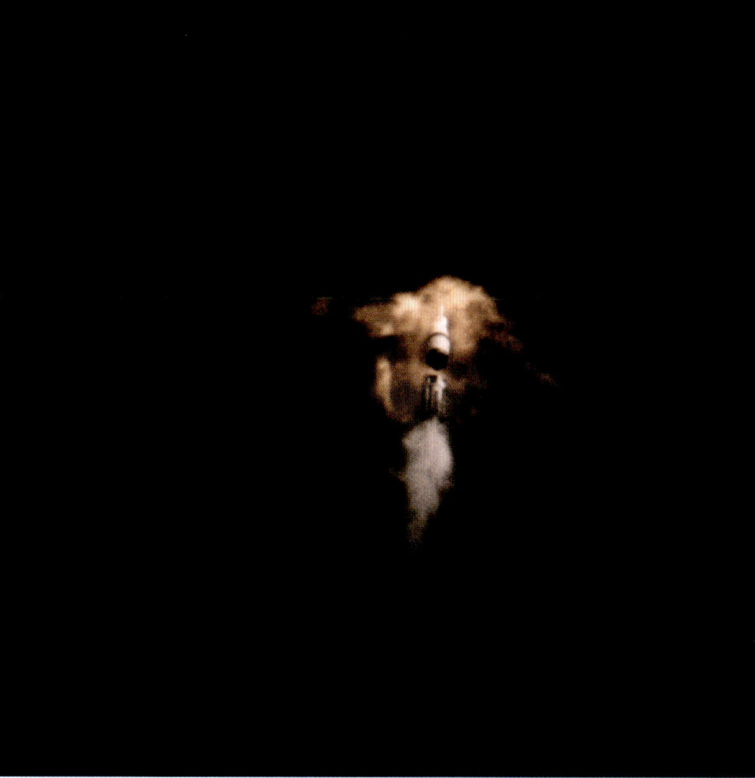

Separation of the first stage at 2:45 into the flight is captured by a ground-based Eastern Test Range Recording Optical Tracking Instrument telescope at Melbourne Beach, Florida, 40 miles south. The vehicle is at 38.5 miles in altitude. Apollo 7 will be in orbit about seven minutes later. Through the Bermuda tracking station, Schirra reports "a little bumpy ride on [the second] stage, but very pleasant."

News photographers follow the ascent. The booster is visible from the ground through first-stage separation and beyond.

Launch guests crane their necks at the booster heads higher. The CSM/S-IVB combination enters a slightly elliptical orbit 176 miles by 141 miles in altitude at ten minutes and thirty-four seconds after launch.

Rockwell (*left*) watches with Kirk.

Von Braun, Phillips, George Mueller, Debus, and Goddard Space Flight Center director John Clark at the postlaunch news conference, with Jack King (*right*) moderating. With no more Saturn IB launches for at least four years, Chrysler announces that about a hundred employees would be let go from its launch team. Debus says on October 15 that pads 34 and 37 will be closed, leading to a reduction in the KSC contractor workforce of 1,315 persons by January. Another one thousand are reassigned to work on the Saturn V. LC-34 and 37 are maintained for the Apollo Applications Program (later named Skylab), which will use Saturn IBs.

CHAPTER 9

October 11–21, 1968

With Apollo 7 in orbit, flight controllers in the Mission Operations Control Room (MOCR) I, on the second floor of the MCC (Building 30) at MSC, are busy. Astronauts Dave Scott (*green shirt, center*) and Rusty Schweikart (*to Scott's right*) gather near the assistant flight director's console.

Lead flight director Glynn Lunney supervises the first shift of controllers after launch with his Black Team. The second shift (White Team) is led by Gene Kranz; the third (overnight) shift (Gold Team) by Gerry Griffin. It's the second use of the new MOCR, first used for Apollo 5.

The CSM moves closer to the S-IVB in this McDonnell Douglas art. Before separating from the stage nearly three hours into the mission, the crewmen manually "fly" the entire combination for a few minutes.

The four open SLA panels reveal the cross-shaped structural stiffener inside the S-IVB, which substitutes for an LM on this mission.

Apollo 7 approaches the S-IVB during the second orbit. The rendezvous is intended in part to simulate rescuing a stranded LM in lunar orbit.

A photograph over Sonora, Mexico, shows that one SLA panel is not open as far as the others. It had initially fully opened to 45 degrees but rebounded a bit due to a cable issue. NASA will simply jettison the panels on future flights to avoid such a problem.

The CSM is only about 5 feet away from the S-IVB during its closest approach. We first see and explain the target on page 102. The white disc (*at lower left*) is a docking target like those that will be used on LMs starting with Apollo 9. A SLA access door is at center.

The CSM begins to move away from the booster.

The stage passes over the Bay of St. Louis and Lake Borgne area just east of New Orleans, Louisiana, about 1:00 p.m. EDT, about 100 feet away from Apollo 7.

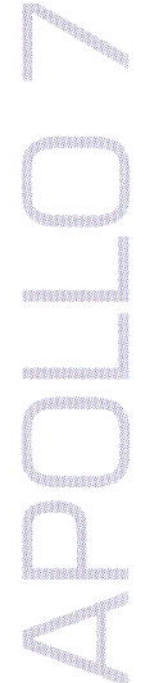

At 1:20 p.m. over Australia, Apollo 7 makes the first of two brief SPS firings to lower its orbit to end up about 90 miles in front of the stage for a second rendezvous attempt the next day. "That beauty really socks it to you," reports Schirra. "That's a real kick in the fanny." It is the most powerful engine ever fired on a manned spacecraft.

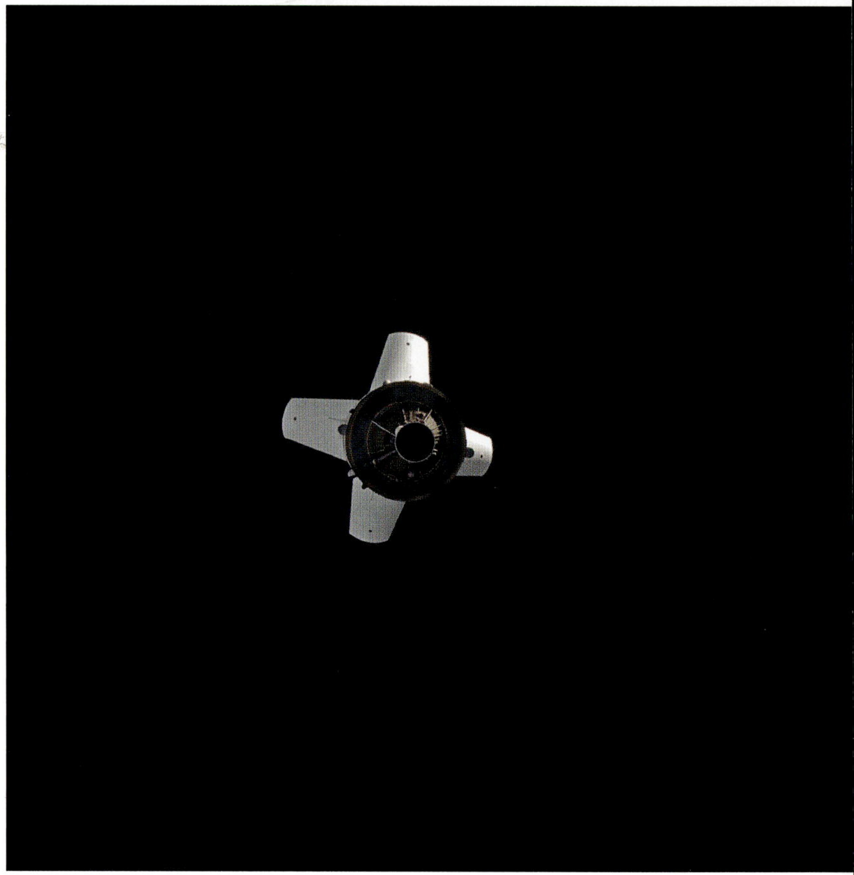

Schirra takes notes before the October 12 rendezvous while looking through the Crewman Optical Alignment Sight to help spot the booster at a distance. It takes more than three hours to close the gap with the S-IVB. An early opponent of having a TV camera aboard, he vetoes that morning's planned first telecast. "You have added two burns to this flight schedule, you have added a urine water dump, and we have a new vehicle up here, and I tell you this flight TV will be delayed without further discussion until after the rendezvous," he tells capcom Jack Swigert. "We do not have the equipment out, we have not had an opportunity to follow setting, we have not eaten at this point, I still have a cold, [and] I refuse to foul up our timelines this way."

The astronauts find that the stage is tumbling "rather wildly" when they meet up with it for the second time.

An aft view of the S-IVB shows its J-2 engine nozzle.

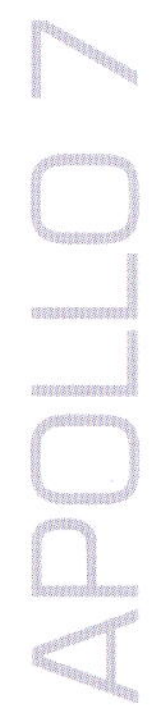

The S-IVB during the second rendezvous. The CSM gets as close as 70 feet and stays there for half an hour.

With the 16 mm Maurer sequence camera (*at left*), Cunningham takes notes during the rendezvous. His Plantronics lightweight headset is around his neck. The CM's hatch handle is at top left.

The second rendezvous with the S-IVB ends at 5:00 p.m. EDT on October 12.

Schirra on October 12. Cunningham reports to Mission Control that the commander has "a pretty bad head cold" and "has been blowing his nose pretty much all day long."

Cunningham on October 12. He and Schirra try to sleep seven hours each night as Eisele monitors the CM before his sleep period starts at 6:00 p.m., a schedule that has all three awake for the TV downlinks.

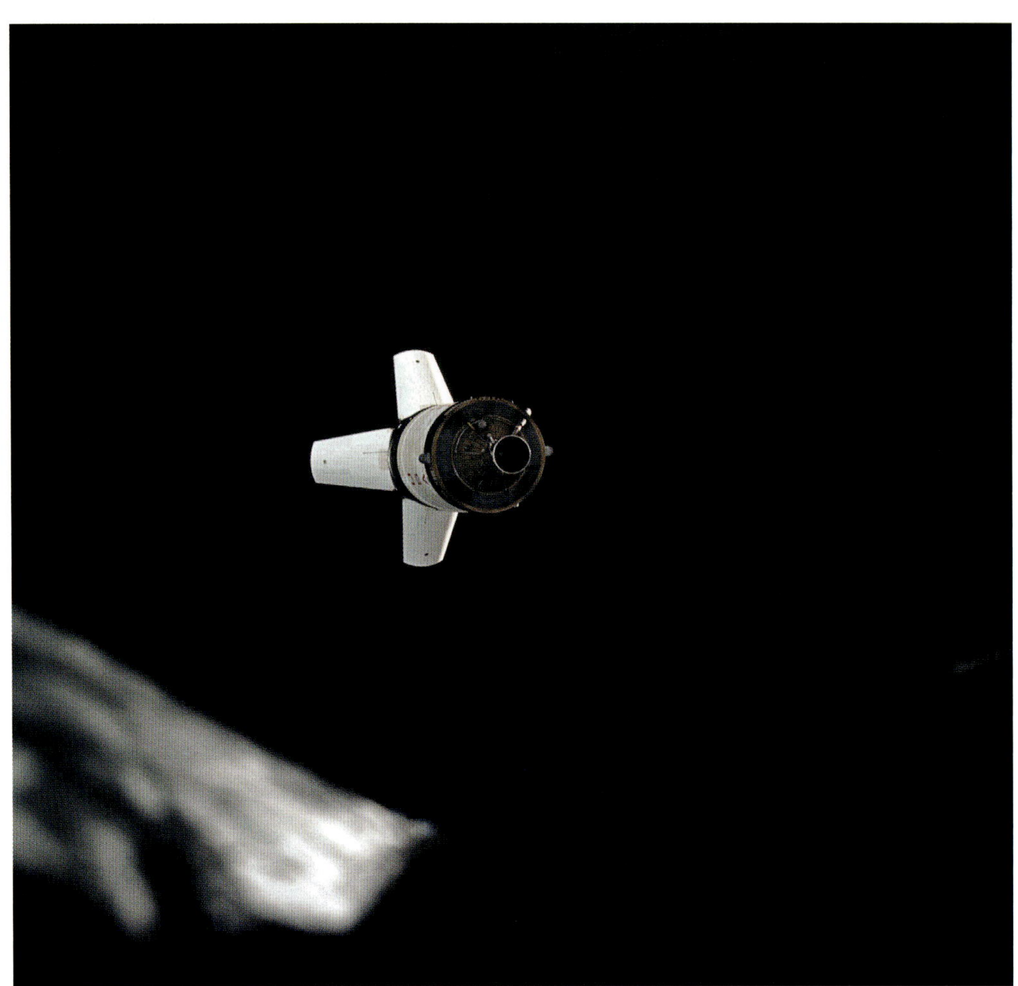

The Southern California coast north of Los Angeles as seen from 125 miles in altitude on October 12. This area is repeatedly photographed under different camera and sun angles.

Left to right: Cunningham's daughter, Kimberly, seven; wife, Lo; and son, Brian, five, arrive outside the MCC on the morning of October 12. They stay for about an hour with a houseguest and family friend, from California.

Lo Cunningham (*left*) joins MSC deputy director James Elms in the MOCR viewing room.

Brian and Kimberly Cunningham gaze through the viewing-room window, with a "squawk box" providing audio.

Wally Schirra had accomplished the first successful space rendezvous on December 15, 1965, as command pilot of Gemini VI, when he maneuvered to within 1 foot of Gemini VII. The spacecraft were not equipped to dock but maintained station keeping for more than twenty minutes. Schirra and pilot Tom Stafford (Apollo 7 backup commander) were also the first to display a humorous sign during a space mission, placing a "Beat Army" placard in their window for the Gemini VII crew to see.

Schirra holds a screw from the instrument panel. *16 mm still frame*

Eisele works with his communications cable in the lower equipment bay, with the sextant behind him (*at left*). The crewmen will report that moving around in the roomy Apollo CM is effortless, and they are surprised by how easily they can maintain their positions without bracing themselves. *16 mm still frame*

Cunningham floats above the crew couches, with the CM's rotational hand controller at center left. *16 mm still frame*

Harriet Eisele (*right*) leaves Seabrook Methodist Church after morning services on October 13 with her parents, Mr. and Mrs. Harry Hamilton, and her son Jon. *AP photo*

The Sinai Peninsula, Red Sea, Gulf of Suez, and Gulf of Aqaba are photographed from an altitude of 147 miles on October 13. Cunningham, who's also suffering from a cold, reports that he and Schirra are feeling better.

Mexico's Yucatán Peninsula is photographed during Apollo 7's thirty-third orbit on October 13, more than fifty-two hours into the mission.

NASA first experimented with television from orbit during Mercury astronaut Gordon Cooper's 1963 flight. He briefly transmitted slow-scan black-and-white TV images to Mercury Control at a rate of one frame every two seconds, with a resolution of 320 lines of horizontal resolution.

For Apollo, RCA built six 4.5-pound vidicon TV cameras that produced ten frames per second at 320 lines (five were refurbished for Block II use). Two lenses were provided: a 160-degree wide-angle lens and a 9-degree-angle lens. The TV signal was multiplexed into the CM's new Unified S-Band system for downlink.

Only two of NASA's Apollo Network tracking stations had equipment to convert the transmissions for the TV networks. The Corpus Christi, Texas, and KSC's Merritt Island stations each used a prototype RCA scan converter built for NASA, with a magnetic disc recorder similar to those used for sports instant replay (KSC's converter was in the MSOB). The slow-scan signal was converted to the US broadcast standard of thirty frames per second and 525 lines of resolution and sent to MSC on landlines, which then released the transmissions to the three national broadcast networks. CBS and NBC carried them live; ABC delayed the first three for about ten minutes.

During the fifth TV downlink, Schirra said the crew was "going to try for an Emmy for the best weekly series." In 1969, the Apollo 7 astronauts received a special Emmy Award from the National Academy of Television Arts and Sciences for their broadcasts.

On the morning of October 14, flight controllers watch the first of seven telecasts from the spacecraft during the next week. "A pretty show for the whole family," says Cunningham.

Eisele (*left*) and Schirra add some humor with their hand-lettered cards. NASA later reports more than three thousand letters arrive at MSC, and five hundred more at KSC.

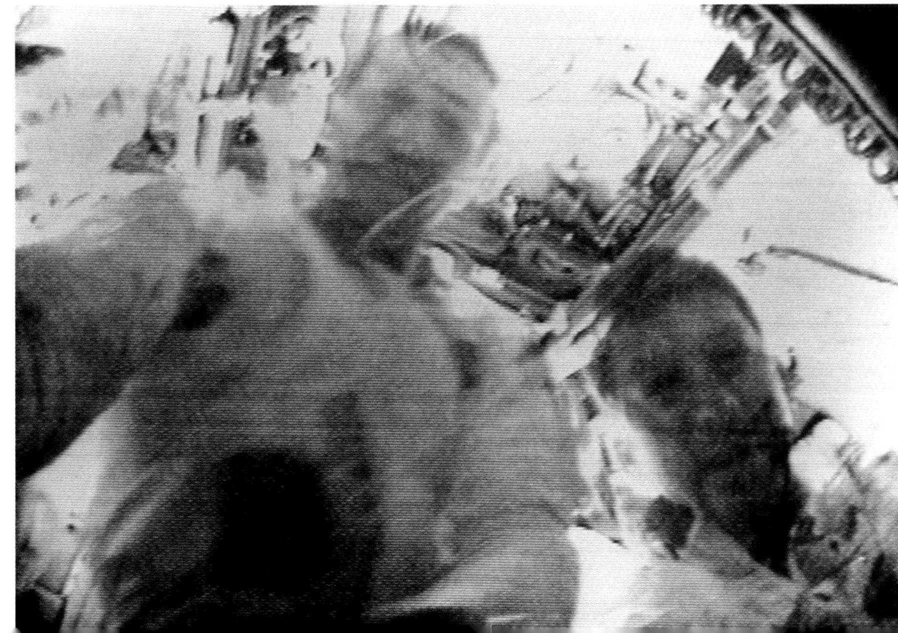

Eisele and Schirra smile during the 10:44 a.m. EDT telecast, which lasts seven minutes during their forty-fifth orbit. The length is limited by the spacecraft's time over the US, where the only two tracking stations that can convert the TV signal are located—KSC and Corpus Christi, Texas.

Flight directors Glynn Lunney (*left*) and Gerry Griffin (*off-going*) in the MOCR on October 14. "[The TV] looks really good; I'm amazed," Capcom Tom Stafford tells the crew.

Jo Schirra (*left*) and Deke Slayton's wife, Marge, in the viewing room on October 14. Mission commentator Paul Haney notes that Jo smokes two cigarettes during the telecast.

Eisele and Schirra during the telecast. Cunningham briefly points the camera out a window to show the Gulf coast.

Jo Schirra shows a printout of a video image outside the MCC. *AP photo*

Cumulus, altocumulus, and cirrus clouds photographed with Earth's limb on October 14

Schirra on October 14. The Apollo 10 astronauts would be the first to shave in space.

Eisele on October 14. He conducts the third SPS burn from the commander's couch to lower Apollo 7's perigee to better conserve propellants. The connector near his cheek is for his headset.

Cunningham reports seeing and tracking a star during daylight by using the sextant, and he and Eisele mark their fifth anniversary with NASA.

At 10:20 the next morning, Eisele holds the "from the Lovely Apollo Room high atop everything" sign, which had opened their first telecast. "Coming to you live from outer space, the one and only original Apollo orbiting road show," he says.

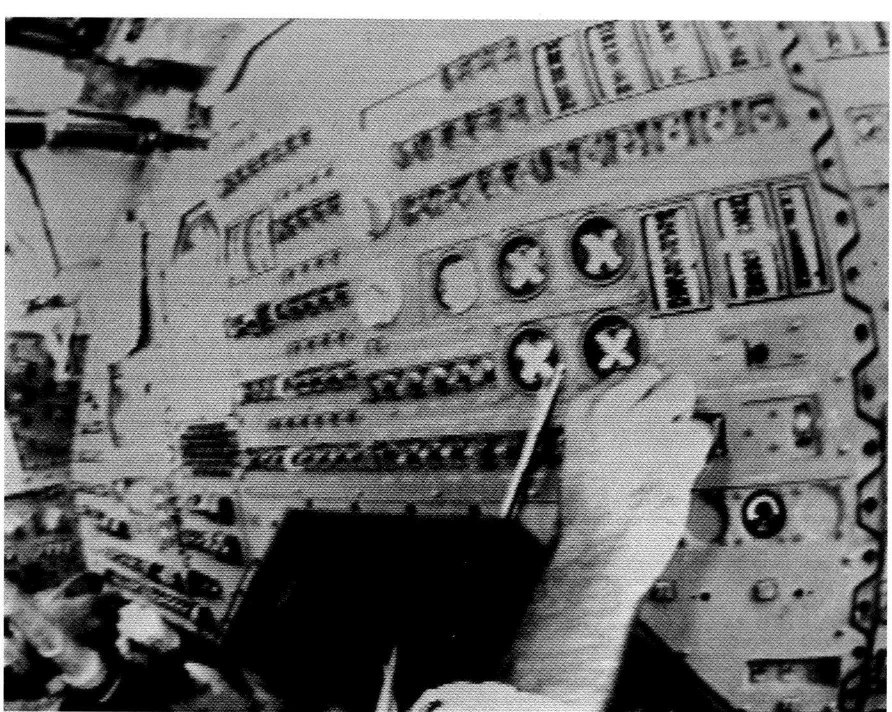

Schirra reaches for a pen floating in front of the main instrument panel during his TV tour of the cabin on October 15.

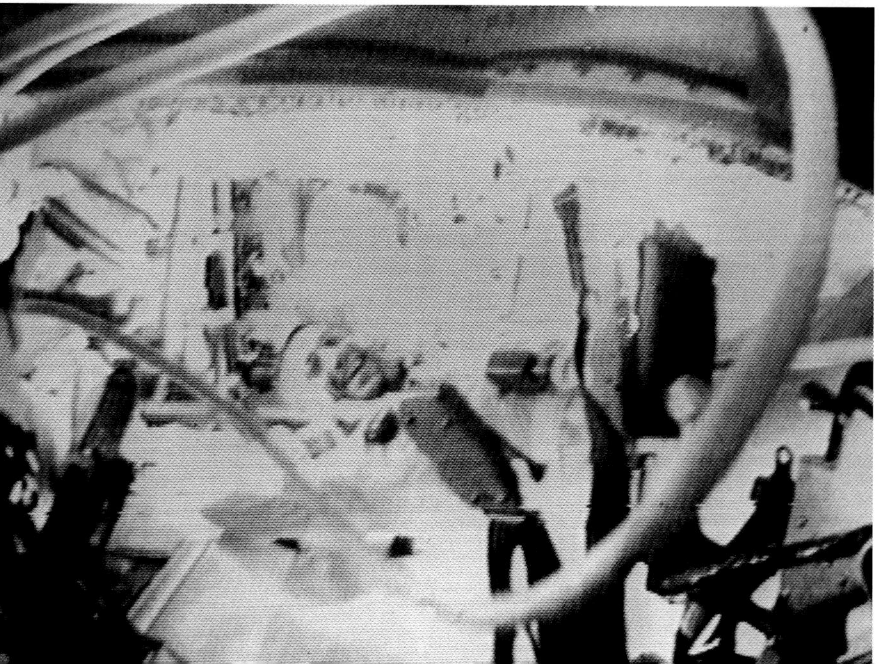

Eisele in the commander's couch during the October 15 telecast

Harriet Eisele watches from the MOCR viewing room during the eleven-minute telecast. Her parents join her.

Hawaii on October 15. The crewmen, who snap more than five hundred photos with their Hasselblad 70 mm camera, report the magazine and frame numbers of each image to flight controllers. Apollo 7's terrain photography was cleared for release only after a team from the CIA and DIA (Defense Intelligence Agency) immediately inspected each frame at MSC to ensure that information vital to the national security of the US and its allies would not be compromised.

The Siling Lake in the Plateau of Tibet in southwestern China is photographed from an altitude of 147 miles on October 15.

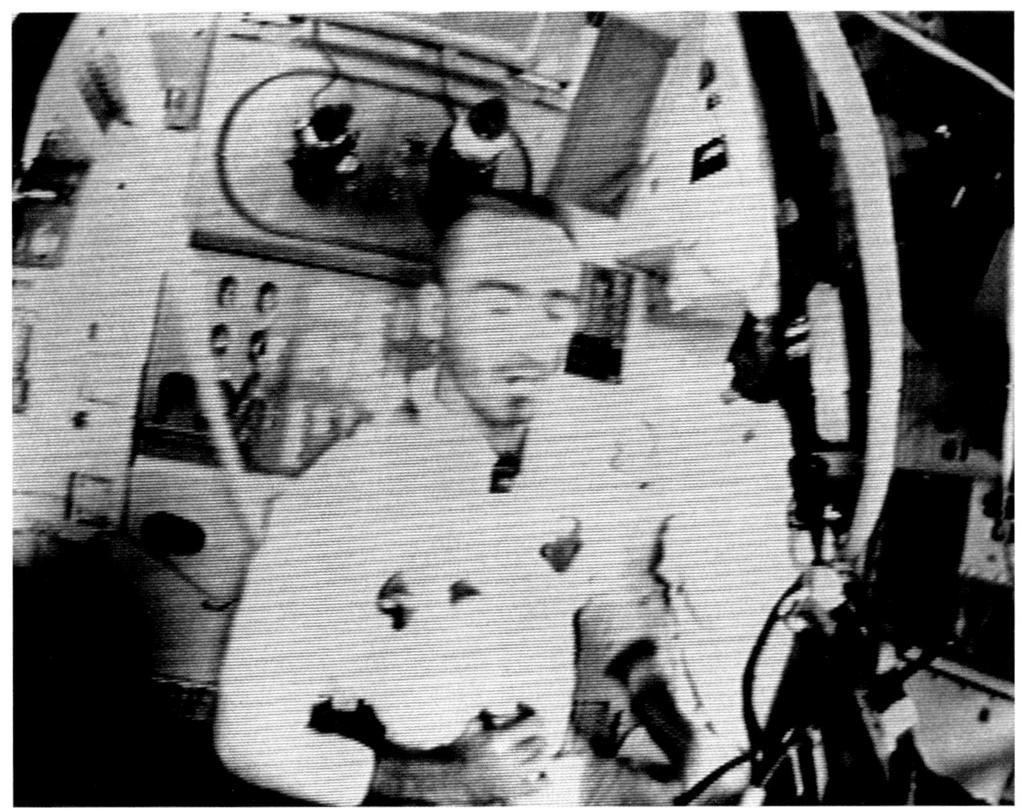

Cunningham, in the lower equipment bay, explains how liquids are vented during the twelve-minute October 16 telecast. The sextant and telescope are above him. Eisele will next show a puddle of water condensation from a cold glycol line; NASA officials had asked the crew to tone down the hijinks.

Cunningham watches Eisele point out the manual RCS switches on October 16. Cunningham's wife, Lo, and his brother Bill and his wife, who were also at the launch visiting from Alaska, are in the viewing room on October 16.

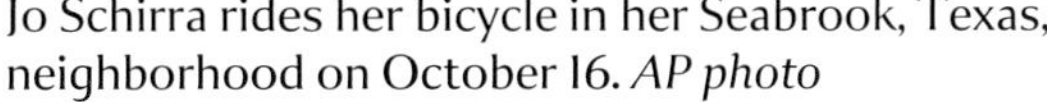

Jo Schirra rides her bicycle in her Seabrook, Texas, neighborhood on October 16. *AP photo*

MSC director of flight operations Chris Kraft (*left*) and George Low in the MOCR on October 17, with the viewing room behind them. Flight director Glynn Lunney (***almost out of frame***) is at lower left. "Were we to reenter now, we would have accomplished 75 percent of all the things we laid out to do," Lunney tells reporters at the flight's halfway point.

During the October 17 telecast**,** Schirra shows the Maurer 16 mm camera, which he says has been recording their "home movies." The crewmen go over sleeping accommodations and show suit and helmet storage bags. Flight controllers judge the video quality of this transmission, which begins at 8:17 a.m., to be subpar.

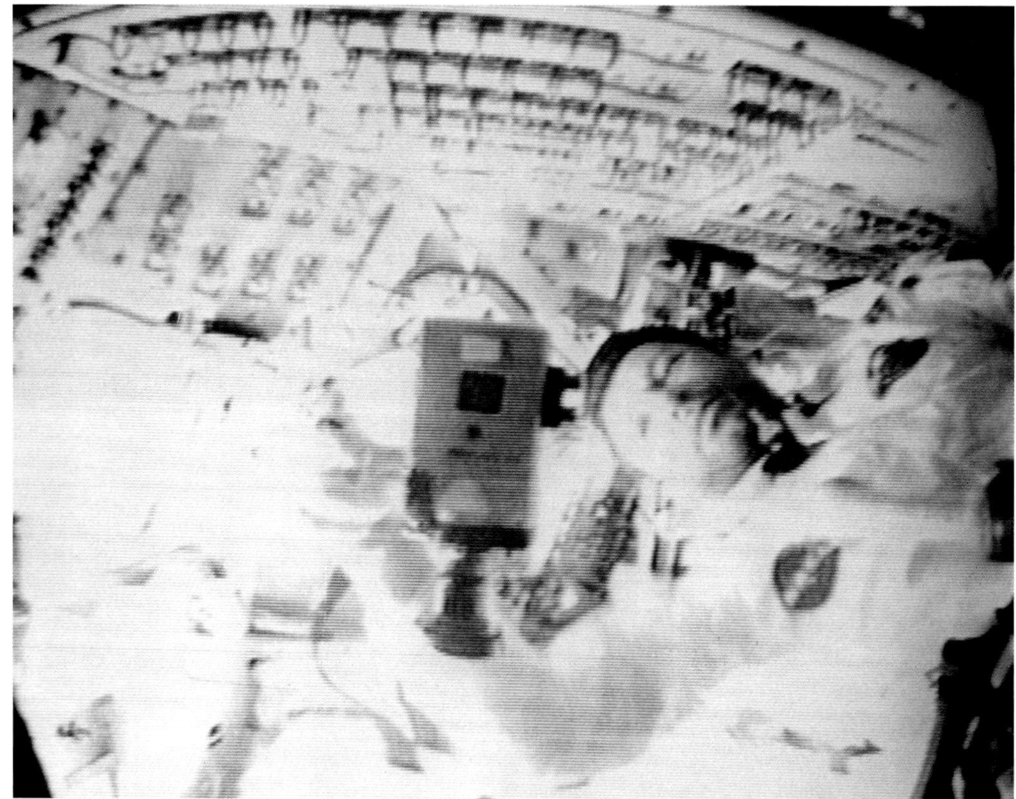

Kraft speaks with Maj. Gen. Vincent Huston, manager for Manned Space Flight Support Operations, as Low listens in the top row of the MOCR on October 17. Huston is the single-point contact between the Defense Department and NASA. Apollo mission director William Schneider is at left.

Hurricane Gladys churns in the Gulf of Mexico during Apollo 7's ninety-first orbit on October 17. "That hurricane is really a doozy," Schirra says during the TV downlink. "I haven't seen anything like that ever." Photographed from an altitude of 114 miles, Gladys is the first Atlantic hurricane to be observed by hurricane hunter aircraft, radar, and space photography. The hurricane had formed in the western Caribbean four days earlier. The crew also photographs Typhoon Gloria in the Pacific.

The spacecraft passes over the southern coast of Sherbro Island on the southwest coast of Sierra Leone at 100 miles in altitude on October 17.

The left side of the forward crew compartment. *16 mm still frame*

The right side of the forward crew compartment. *16 mm still frame*

View of the hatch, looking up from the center couch. *16 mm still frame*

Left to right: Jo Schirra, Lo Cunningham, and Harriet Eisele try to get a glimpse of Apollo 7 passing overhead at a boat landing on Clear Lake just east of MSC early on October 18, but clouds interfere. It was visible over Southern California the previous morning, and it's the first opportunity to see the spacecraft in darkness over Texas. The astronauts cancel the morning's TV downlink, saying they'll be too busy with another SPS burn. *AP photo*

The 30-foot Unified S-Band antenna used to receive TV transmissions at the Apollo Network's KSC station

Bendix field engineer Hugh Huet monitors the S-Band video tape recorder at KSC.

The east coast of Africa and Madagascar on October 18

Bermuda on October 18. The S-IVB burns up during reentry that day, and debris falls into the Indian Ocean.

Cunningham during the twelve-minute October 19 telecast, which starts just before 10:00 a.m. Schirra claims he puts his crew through a daily marching drill, and the astronauts demonstrate as best as they can.

Eisele (*foreground*) and Cunningham on October 19. The crewmen report continuing sinus congestion.

Schirra during the October 19 telecast. He later requests that the crew not wear helmets during reentry, to prevent possible eardrum damage from congestion, but Slayton stands firm that they would. The astronauts also complain about flight plan changes, equipment problems, and lack of sleep.

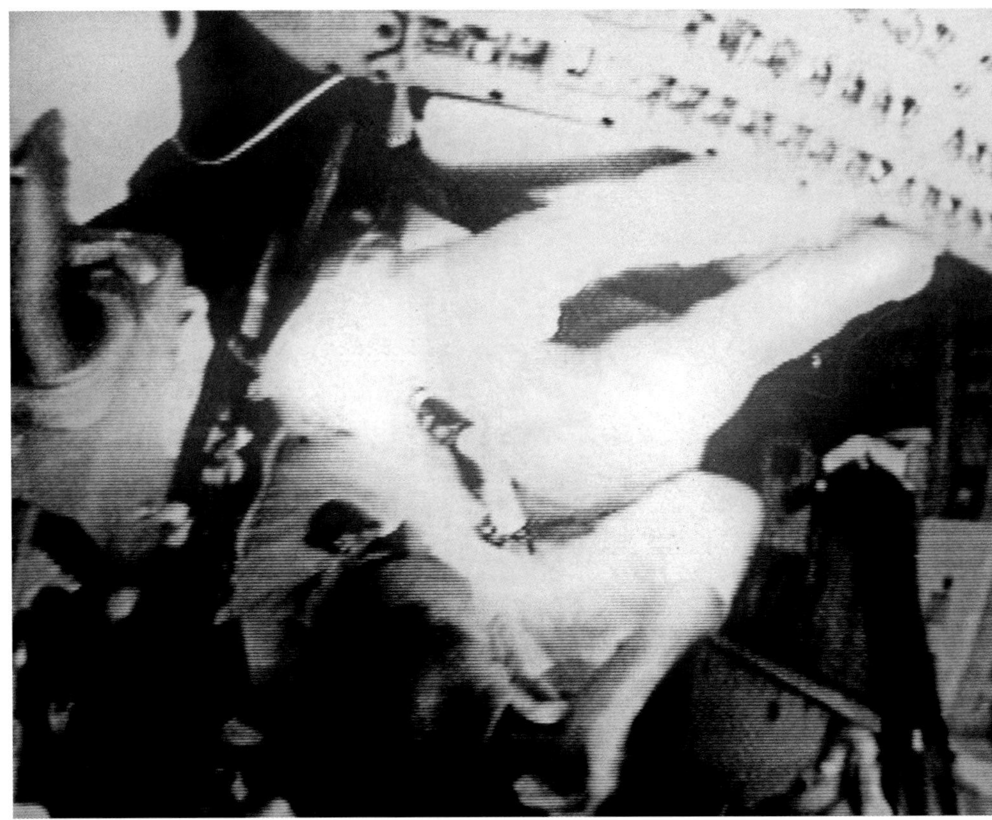

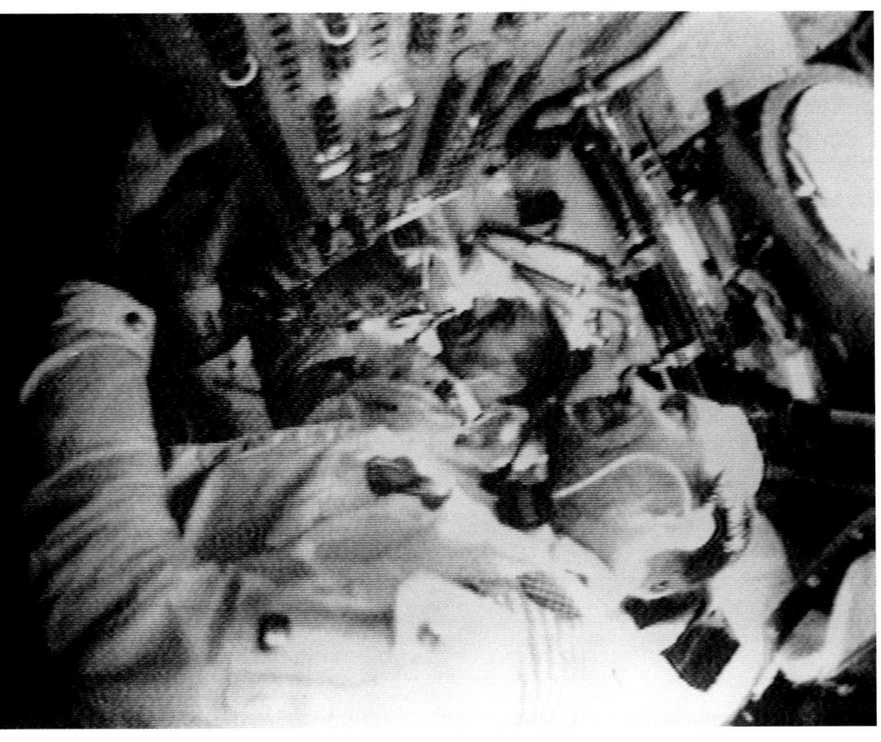

The northeast coast of Australia and the Great Barrier Reef, photographed on October 19

The morning sun reflects off the Gulf of Mexico and the Atlantic Ocean at an altitude of 138 miles on October 20, during the spacecraft's 134th orbit. "That's one of the most spectacular sights I've seen just now," Schirra says during the TV downlink. "We can see the whole Florida peninsula lit up by the sun's rays."

Eisele during their nine-minute October 20 telecast. "They don't let me up here very often," he says as he floats up from the lower equipment bay. "Only for the show."

Cunningham takes notes on October 20 as a Hasselblad camera magazine floats in front of him.

Cunningham (*left*) and Eisele with a flight document on October 20. Later, Schirra torches a just-added experiment. "I wish you would find the idiot's name who thought up this test," he snaps. "I've had it up to here today." He adds, "We're going to be kind of hard-nosed up here from now on."

Left to right: Harriet Eisele, Lo Cunningham, and Jo Schirra attend a Houston Oilers NFL game at the Astrodome on October 20. The Oilers lost to the New York Jets 20–14.
UPI photo

Cunningham with an orange drink pack; he'd used the water gun to rehydrate it. The crewmen report the number of squirts and the time to Mission Control. Hot water is also available for the first time. *16 mm still frame*

Eisele in the lower equipment bay. *16 mm still frame*

Cunningham easily dons his space suit in about eight minutes, a more time-consuming task on the ground. Schirra tells controllers that his crew will wear their space suits but not helmets for reentry: "Our heads are still too stuffed up." *16 mm still frame*

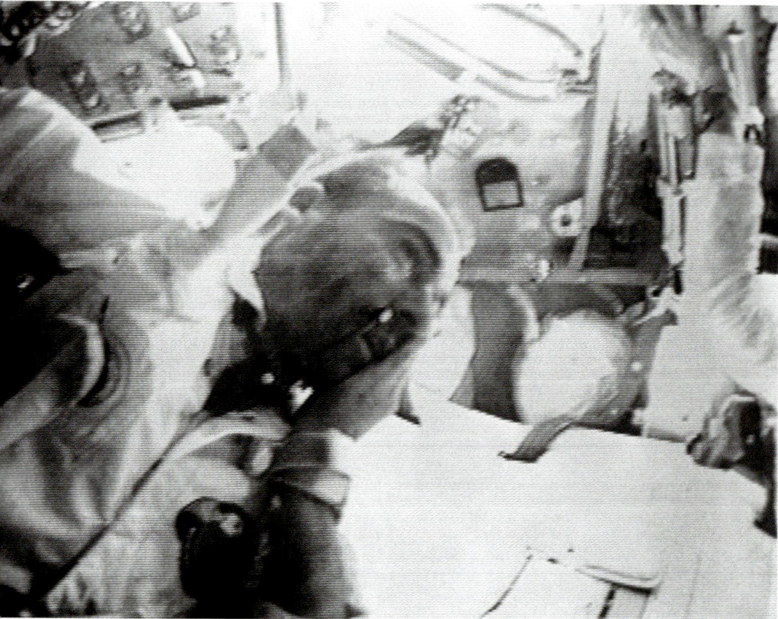

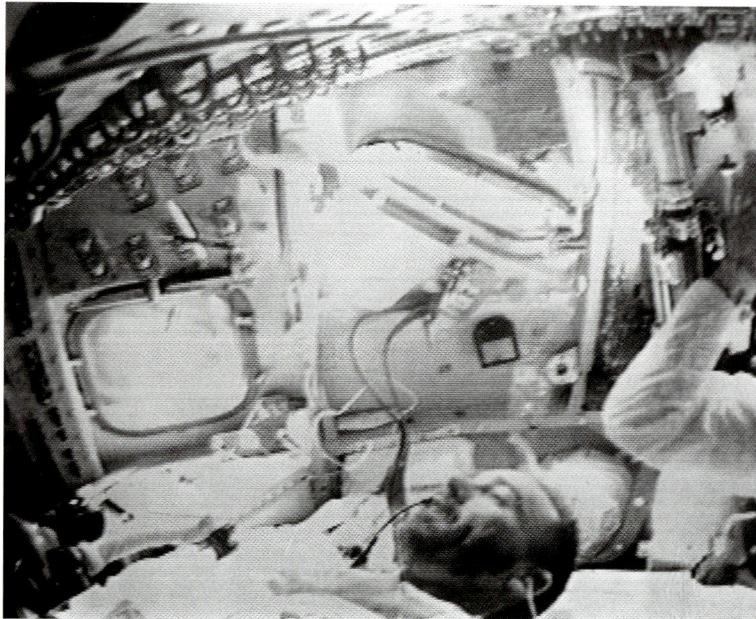

Cunningham during the nine-minute October 21 telecast. The astronauts decide to give viewers close-ups of their beards. "We are not fans of the beard club," Schirra advises.

Cunningham on October 21 as the astronauts prepare for their seventh SPS burn. All three crewmen take decongestant pills every eight hours.

The crew holds a concluding TV card. "This is Apollo 7 signing off," says Schirra to end their seventh and final downlink, which includes showing photos of their wives taped to the control panel.

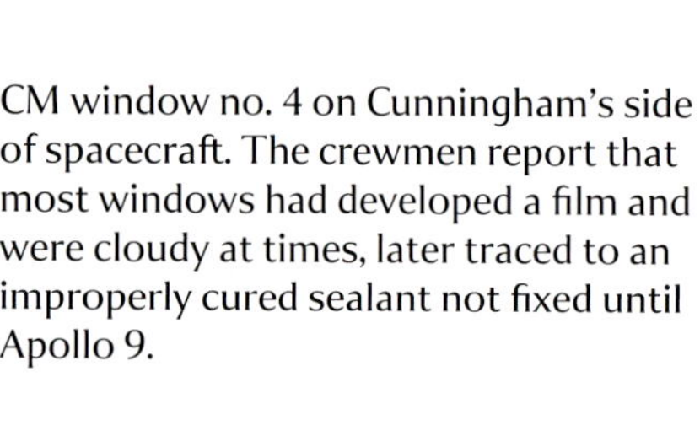

CM window no. 4 on Cunningham's side of spacecraft. The crewmen report that most windows had developed a film and were cloudy at times, later traced to an improperly cured sealant not fixed until Apollo 9.

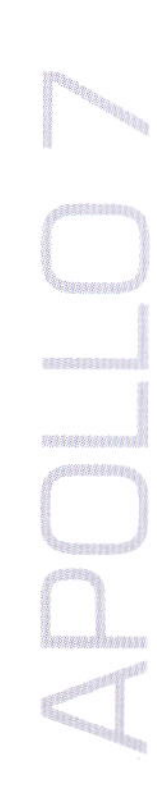

Schirra, in his left-hand couch, looks through a rendezvous window. Shortly before reentry, he tells flight director Kranz, "We want to thank you and the team for an outstanding job. It was a real professional show and one we really enjoyed."

Eisele is happy his first space mission is almost over after eleven days. Although he remained with NASA until 1972, he did not fly again.

CHAPTER 10

October 22, 1968

The Apollo 7 prime recovery ship, the aircraft carrier USS *Essex* (CVS-9), is photographed by NASA senior photographer Bill Taub as he arrives on October 15 for the recovery in the Atlantic Ocean, south of Bermuda. *Bill Taub archive*

The conning tower and bridge of *Essex* bristles with antennas.
Bill Taub archive

Components of Helicopter Squadron 5 wait on the portside deck-edge elevator of *Essex*. The portable network TV uplink is under an inflatable cover (*at lower right*). *Bill Taub archive*

Helo 55 (*background*) will recover the astronauts after splashdown and bring them to the ship. The Navy Sikorsky SH-3A Sea King helicopters are used primarily for antisubmarine warfare. *Bill Taub archive*

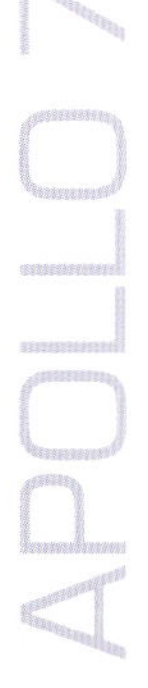

As a colleague naps, an NBC engineer checks out his GE PE 250 camera aboard *Essex* for the first live color TV coverage of a spacecraft splashdown and recovery. *Bill Taub archive*

NBC serves as the TV network pool producer for Apollo 7 and hires Western Union International (WUI) to handle transmission for the first live color coverage of a splashdown. WUI receives special authority from the Federal Communications Commission to operate a TV transmitting facility aboard a US Navy ship and leases a portable General Electric uplink for the deck of USS *Essex*. It sends the TV signal to NASA's ATS-3 satellite over Brazil, which relays it to a new Comsat ground station in Etam, West Virginia. The signal is then sent to New York City on landlines for distribution to the three TV networks. NBC uses five color cameras—four studio cameras and one handheld—and two video tape recorders aboard the ship. The overcast weather, however, prevents the audience from seeing much of anything until the astronauts arrive on *Essex*, who then disappear below deck after just a few minutes, making for fairly uneventful viewing. Schirra's previous splashdown, Gemini VI, was the first to have live TV coverage, also using a shipboard uplink.

Helo 55 at sunset.
Bill Taub archive

The CM is depicted during reentry through the atmosphere at 7:00 a.m. EDT in this NAA art, its heat shield enduring temperatures of up to 5,000 degrees Fahrenheit. The SPS engine had been fired for the eighth and final time for the deorbit maneuver at 6:42 a.m. EDT, a nearly twelve-second burst southeast of Hawaii during the 163rd orbit. The SM is jettisoned four minutes later.

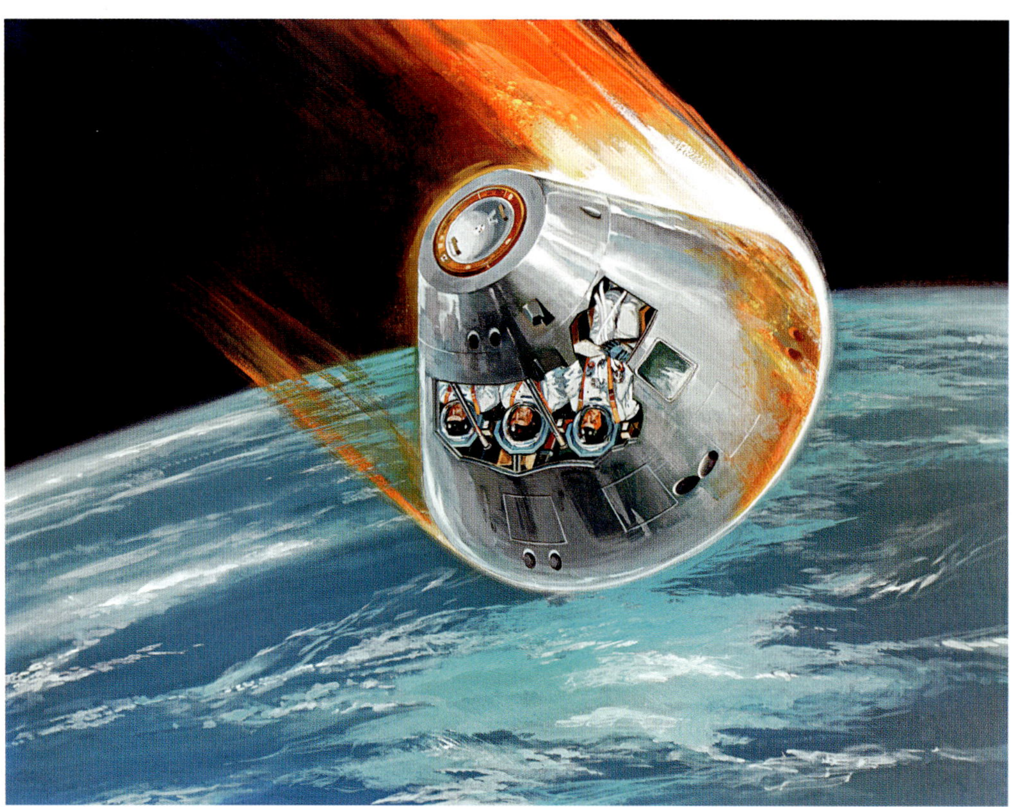

Aerial view of *Essex* on October 22. The TV uplink is above the "7." The ship had also been designated as the prime recovery ship for Apollo 1.

Sailors gather along the ship's railings above a welcome banner.

Flight director Gene Kranz monitors the final minutes of the mission in the MOCR. Splashdown is in the Atlantic Ocean, 300 miles south of Bermuda, at 7:11 a.m. after ten days, twenty hours, and nine minutes, but the crew is out of communication for fifteen minutes because waves tip the CM to its the apex-down stable II position, submerging its homing beacon.

The CM is righted within seven minutes by the inflatable uprighting bags, and divers secure the flotation collar, dropped from Helo 55, designated Recovery Helo 3 (*left*), at 7:30 a.m. One diver jokes that Schirra, a Navy captain, is getting his first submarine duty. The impact point is only about 1,700 feet from the target, but *Essex* is more than 7 miles away.

Left to right: Director of flight operations Chris Kraft, MSC director Robert Gilruth, deputy Apollo program director George Hage, Apollo program director Sam Phillips, and Apollo Spacecraft Program manager George Low celebrate with cigars in the MOCR after the splashdown.

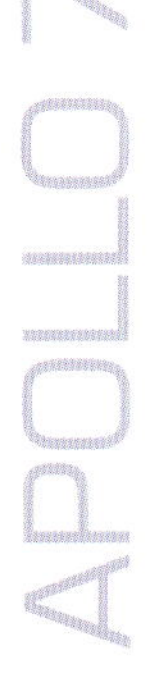

With two life rafts inflated, the divers prepare to open the new one-piece hatch at 7:48 a.m. The astronauts have removed their space suits and donned their constant-wear garments. *Essex* is now less than a mile away.

Divers help the astronauts into a waiting raft. Eisele is the first to egress, followed by Cunningham. The weather is overcast and rainy but warm.

Schirra is the last to emerge at 7:54 a.m. Areas of the silver reflective tape covering the CM have been burned away.

Divers help Schirra egress.

Schirra is lifted to Recovery Helo 3 in a Billy Pugh rescue net. "Their physical status is good," the chopper radios. "All are in good shape."

In the MOCR, George Mueller, the NASA associate administrator for space flight, and Phillips are pleased with the mission. They are expected to soon officially sign off on sending Apollo 8 around the Moon in December.

The crew emerges from the helicopter on the rain-swept *Essex* flight deck at 8:09 a.m., arriving fifty-six minutes after splashdown. NASA photographer Bill Taub is front and center between two sailors.

The astronauts pause briefly to steady themselves, with Donald Stullken, NASA recovery team leader from MSC's Landing and Recovery Division, at right. He had invented the CM's flotation collar.

[10-021] Schirra surveys the scene with his crewmates as a band strikes up "Anchors Away." Eisele holds flight documents. Schirra had borrowed a crewman's comb before stepping out.

Left to right: Schirra, Cunningham, and Eisele are happy. "It's great to be back," Schirra says. "The mission went beautifully."

Left to right: Flight directors Gene Kranz, Glynn Lunney, and Gerry Griffin pose in the MOCR. Slayton is over Kranz's shoulder.

Jo Schirra, Hariet Eisele, and Lo Cunningham each watch the arrival at home. "It looks like they've lost weight," says Mrs. Schirra (*left*) in Timber Cove. "I hope he isn't too attached to [his beard]." In Nassau Bay, Mrs. Cunningham (*right*) has another view: "Those beards don't look bad at all . . . he went to work, and he just came back." In El Lago, Mrs. Eisele notes that "[Schirra] said [the CM] was a lousy boat." All three later spoke by phone to their husbands at the MCC. *UPI photo*

Stullken points the crew toward the brass. MSC flight surgeon Dr. Clarence Jernigan is behind Stullken.

Cunningham (*left*) waits to follow Eisele and Schirra.

Schirra and Stullken lead the crew as Dr. Jernigan steadies Cunningham (*left*).

Schirra greets RAdm. Thomas Davies, commander of Carrier Division 20, after saluting.

Cunningham greets *Essex*'s captain, John Harkins, with RAdm. Davies (*at left*).

Left to right: Schirra, Eisele, and Cunningham, with Stullken, Dr. Jernigan, and Capt. Harkins behind them

The crewmen briefly pose for photographers, with RAdm. Davies and his chief of staff, Capt. Richard Palkovic, in the background.

Schirra on deck with Stullken, who sports an *Essex* tie clip (*left*)

Eisele with Dr. Jernigan (*left*)

Cunningham removes his sunglasses and squints in the morning sunlight.

Stullken leads the crew toward the deck-edge elevator.

Left to right: Cunningham, Schirra, and Eisele prepare to take elevator no. 2 from the flight deck. At right is Dr. Bill Carpentier, the recovery team's chief physician.

Left to right: Cunningham, Eisele, and Schirra quickly descend to the hangar deck.

The group heads below.

Cunningham (*left*) and Dr. Jernigan walk toward the ship's sick bay for the crewmen's medical examinations. Their eardrums will show no signs of damage.

The astronauts and Stullken arrive on the hangar deck.

President Johnson congratulates the astronauts from the Cabinet Room at the White House at 8:53 a.m., after their steak-and-eggs breakfast. "We . . . are so very proud of you this morning," he says. "When you have finished your debriefings, Mrs. Johnson and I hope to receive you where we can talk about your experiences without having to go through the Houston switchboard." He watches the first live color recovery coverage on a local Washington, DC, TV station. *Photo by Mike Geissinger / LBJ Library*

Left to right: Eisele, Schirra, and Cunningham each reply, "Thank you very much, Mr. President," with Cunningham adding, "It was our pleasure and honor to make the trip." They then head for the showers and a shave.

Jo Schirra and daughter, Suzanne (*left*), talk with reporters outside their home. Suzanne, asked about the men's beards, comments that "Gee, after they shave, they're going to be real people again." Her mother has changed clothes since the splashdown. *Photo by AP*

Divers attach a recovery loop from *Essex* to Apollo 7 about two hours after splashdown.

Divers work with tie lines before bringing the CM aboard.

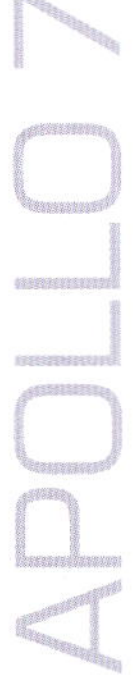

The CM is carefully hoisted from the ocean. Before its retrieval, an auxiliary recovery loop had been attached to the spacecraft for safety. Its cable was taped to the recovery loop before hoisting, and the size of the resulting cable makes it hard for the ship to hook on.

The flotation collar falls from the CM as the ship's crane swings it over the railing. The collar had not fit properly after inflation, which will trigger an investigation.

Sailors help lower the CM onto a dolly in the foreground.

The heat shield shows the offset ablation pattern from the heat of reentry.

After freshening up, Schirra, Cunningham, and Eisele take a walk on the *Essex* deck.

Left to right: Cunningham, Schirra, and Eisele pose in front of the welcome banner. As is not uncommon, Donn Eisele's first name is misspelled.

The astronauts pose on the deck, with Stullken (*at right*).

On the hangar deck, Cunningham and Schirra (*right*) peer into their spacecraft, with Eisele to the left. The CM's windows, which became cloudy at times during the mission, are covered to preserve their condition for study.

Left to right: Cunningham, Schirra, and Eisele pose with Apollo 7.

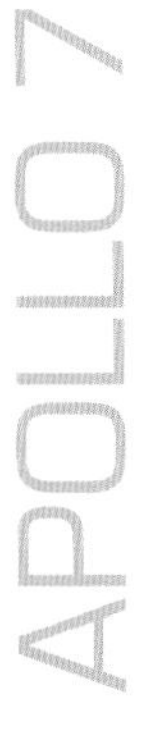

Left to right: Stullken, Schirra, Cunningham, and Eisele on the hangar deck

The CM's hatch pump handle bears a red "DO NOT [REMOVE]" streamer. The drogue parachute mortar cannisters near the top are empty.

The interior of the crew cabin. The astronauts had taped extra clothing to their couch headrests before they reentered without helmets. “The three men who occupied it for eleven days must be awfully good housekeepers,” says Stullken.

NASA technicians secure Apollo 7’s hatch on October 22. The CM will be offloaded from USS *Essex* on October 24 at Norfolk Naval Air Station in Norfolk, Virginia, and flown in a Douglas C-133B Cargomaster to Long Beach, California. It will then be trucked to NAR in Downey.

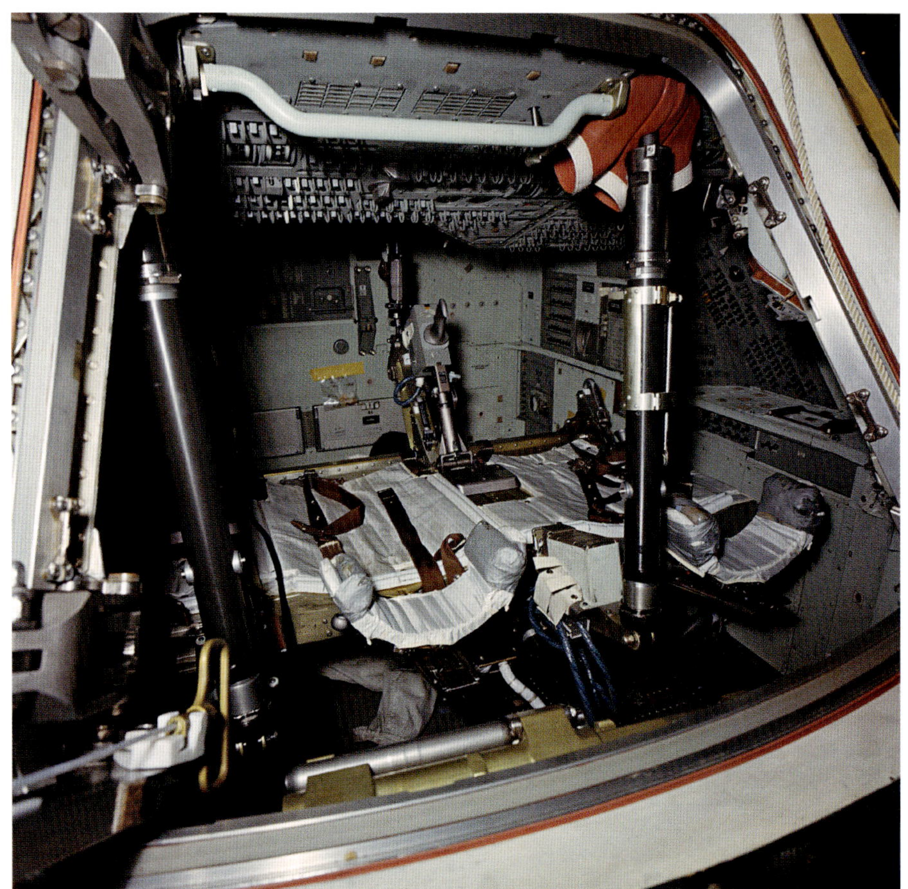

RAdm. Davies sits between Cunningham and Schirra during the first course of a steak dinner in the wardroom.

Cunningham, RAdm. Davies, Schirra, Cmdr. David Carruth, and Eisele during dinner

Eisele chats with Capt. Harkins, with Cmdr. Carruth, the ship's executive officer, to the left.

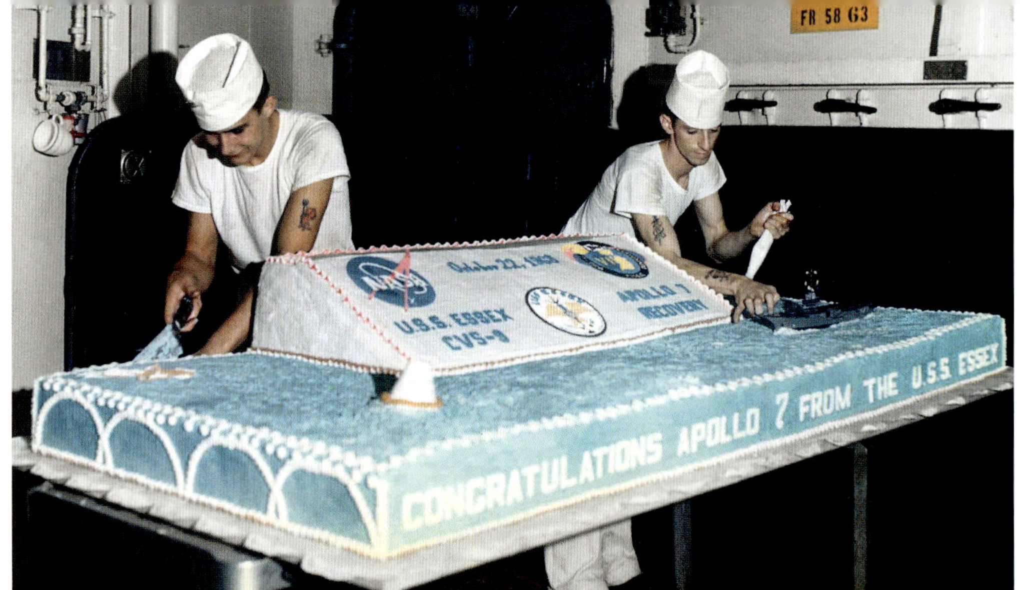

Essex bakers apply the finishing touches to the traditional recovery cake in the galley before rolling it to the ceremony.

The bakers pose with their 700-pound creation, which includes an *Essex* model (*at right*) and a CM (*at left*) in the Atlantic.

Cmdr. Carruth (*second from right*) and Capt. Harkins prepare to honor Cunningham, Eisele, and Schirra on the hangar deck after dinner. “It is a pleasure to be back,” Schirra says. “It looks like the Atlantic Fleet has got me for real.”

The astronauts receive personalized CVS-9 caps from Cmdr. Carruth during the brief ceremony. Schirra then swears in three sailors who are reenlisting for four more years.

The crewmen thank the galley crew for the cake.

Schirra makes the first cuts with a ceremonial sword, and more than a thousand sailors line up for a piece. The astronauts later munch on popcorn while watching a replay of their TV transmissions, before retiring to the captain's quarters for the night.

CHAPTER 11

October 23–December 9, 1968

The crew has breakfast in the captain's dining room aboard *Essex* on the morning of October 23. *Left to right*: Cunningham, Capt. Harkins, Schirra, and Capt. Palkovic.

RAdm. Davies chats with Schirra (*left*), with Eisele, Stullken, and Capt. Palkovic.

Cunningham comes onto the flight deck for departure, with Capt. Harkins (*at left*).

Left to right: Stullken, Cunningham, Capt. Harkins, RAdm. Davies, and Schirra

Left to right: Cunningham, Capt. Harkins, Schirra, Stullken, Eisele, and Cmdr. Carruth on the flight deck

Capt. Harkins helps Eisele with his life jacket (*left*) as Cmdr. Carruth talks with Cunningham (*right*). The two astronauts will depart first in a Grumman C-1A Trader, with Cunningham serving as copilot.

Schirra inspects a hydraulic leak that developed in one wing almost immediately after his C-1A departed, forcing a return to the carrier. "Sorry about that, sir," an officer tells him as he changes planes. "That's all right," Schirra replies. "I've got to get my flying time in somehow."

Schirra's C-1A is finally ready for takeoff for the 420-mile, two-hour-and-fifteen-minute trip to Cape Kennedy. The plane carrying Cunningham and Eisele has been circling, so both aircraft can arrive together.

Schirra (*right*) serves as copilot during the flight, with Navy lieutenant Jeff Walker as pilot.

At the Cape Kennedy AFS Skid Strip, an NBC pool cameraman gets ready for the crew arrival; the GE PE-250 camera is on loan from WTVT-TV in Tampa, Florida. The control tower sign notes that the facility has an elevation of 10 feet above sea level.

A band from Patrick AFB entertains about five hundred NASA and contractor employees waiting for the astronauts to arrive.

KSC director Kurt Debus warms up the crowd before the Apollo 7 crewmen land.

Schirra waves from the C-1A cockpit after landing shortly before noon.

Slayton (*left*) welcomes Schirra. "You old son of a gun, it went real well," Schirra tells him. KSC security chief Charles Buckley is at right.

Schirra and Cunningham (*right*) speak with Slayton after the second aircraft lands.

Cunningham, Eisele, and Schirra are welcomed by (*left to right*) Maj. Gen David Jones, commander of the USAF Eastern Test Range; Debus; Rocco Petrone, KSC director of launch operations; and Ray Clark, KSC director of technical support.

Schirra is reunited with Apollo 7 support astronaut Jack Swigert (*right*); the other two support crew members, Ron Evans and Bill Pogue, are also present.

Schirra reaches the end of the line, with Pogue mostly out of frame (*at right*).

Schirra speaks first. “The best part, of course, is to come home,” he says. “We were here for a number of months, but we left rather suddenly.” When a group of girls in the crowd shouts, “Over here, over here!,” Schirra quips, “We’d like to go over there, but we just don’t have that much courage.”

“It’s great to be back here,” Eisele says, “and I want to thank you all for coming out to give us this warm welcome.”

“I don’t know what I expected when I landed here, but I certainly wasn’t expecting this,” says Cunningham. “All I can say is thank you for the warm welcome.”

The astronauts pose for photographers before leaving for the MSOB about 12:10 p.m. Slayton stands at left.

A Pontiac limousine used for visiting dignitaries will take Schirra and his crew to KSC for four days of debriefings and medical tests.

Cunningham and Schirra (*right*) ride in the rear seat, with Eisele and Slayton in front of them.

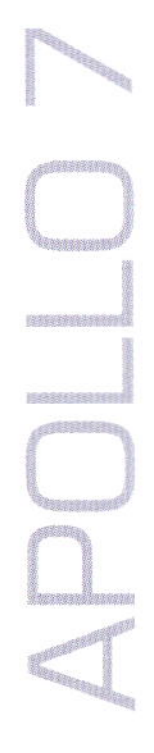

Cunningham waves to a cheering crowd of KSC workers, with Schirra and Slayton at center; Eisele gets out on the opposite side.

Schirra (*left*) heads into the MSOB as Eisele shakes hands with building supervisor Tony Broadway (*hidden by Schirra*). Dearmond Matthews of KSC security is behind Eisele.

Schirra returns to the MSOB, where his crew had departed for their flight twelve days earlier. Buckley is at left, with Matthews at right.

Schirra, Eisele, and Cunningham pose with a sign at the astronaut quarters in the MSOB that poses a classic pilot's question.

Astronaut secretary Lola Morrow (*left*) greets Schirra, as Eisele is welcomed back by Navy lieutenant Dee O'Hara, astronaut nurse. A foil-covered Fisher Price toy TV on the table is one of the gag gifts for the astronauts. A stack of three turtles is on the table (*at left*).

The Ancient and Honorable Order of Turtles started as an informal drinking club among American pilots in England during World War II. When asked by a member, "Are you a turtle?," the required member's public response must be "You bet your sweet ass I am!" to avoid buying the questioner a drink.

During his 1962 Mercury flight, Schirra was the first astronaut to be challenged by the question in space (he replied to Slayton by recording "Y-B-Y-S-A-I-A"). During Apollo 7, Schirra sought revenge during a live TV downlink, with small signs posing the question to Slayton and mission commentator Paul Haney. Pilot and TRW employee Gerry Morton later established the "Interstellar Association of Turtles (Outer Shell Division)," with Schirra as the "High Potentate."

Eisele presents a bouquet of mostly dead roses to Schirra, with Cunningham (*at right*).

Schirra holds a "three wise monkeys" carving presented by Slayton, with KSC mission support chief Hal Collins behind them. Schirra had compared his crew getting around the cabin in orbit to monkeys in a cage.

Cunningham receives his carving. The "Nelson Ravings" sign at right parodies TV's Nielsen ratings, claiming that the mission had more viewers than Dean Martin, astronaut portrayers Bill Dana and Don Knotts, the network news, and Lassie.

Slayton presents Eisele with a carving as Collins looks on. Slayton and Collins wear "Apollo 7 YBYSAIA" buttons.

Schirra has a gag gift for Slayton from the crew.

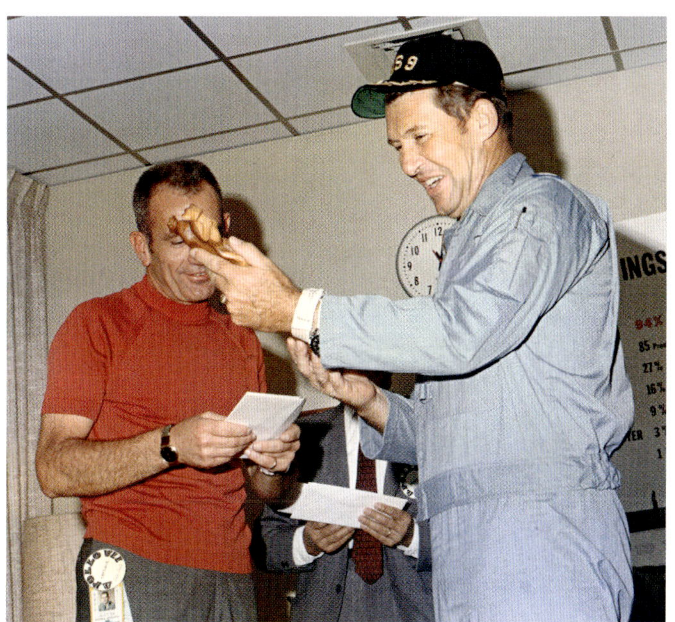

Left to right: Eisele, Schirra, and Cunningham relax in the crew quarters with their gag TV.

Later that afternoon, Cunningham, Eisele, and Schirra begin their debriefings. The next day, they meet with the Apollo 8 and 9 astronauts and their backups.

On October 26, Schirra, Cunningham, and Eisele return to Ellington AFB, southeast of MSC, about 3:30 p.m. CDT aboard a NASA Grumman Gulfstream I. Cunningham greets his daughter, Kimberly, as Eisele's oldest son, Donn H., approaches at right.

Schirra greets his daughter, Suzanne, and wife, Jo; Cunningham hugs his two children, Kimberly and Brian. Also at Ellington to meet the crewmen are two of their backups, Tom Stafford and John Young, and astronauts Alan Shepard and Gordon Cooper.

“It’s been a long trip,” Eisele tells the crowd. “But now we’re back on the ground and finally home.” All three men wear a gold pin of the US submarine service, a nod to the CM being briefly submerged after splashdown. They have no comment on the mission of Soyuz 2, launched on October 26, carrying a single cosmonaut who would fail to dock with an unmanned craft, a technique the Soviet Union had yet to achieve.

Lo, Kimberly, Walt, and Brian Cunningham depart after the brief ceremony.

Donn and Harriet Eisele smile at well-wishers. The crewmen will undergo another week of debriefings at MSC but next head to Schirra's house for a champagne party.

On October 28, the Apollo 7 CM arrives at Building 247 at NAR in Downey, California, for a series of inspections and analysis.

On November 2, the astronauts are guests at the LBJ Ranch in Stonewall, Texas, where President Johnson presents them with NASA Exceptional Service Medals at 10:30 a.m. CDT. Schirra speaks after receiving an oak leaf cluster to go with his previous Exceptional Service Medal for the Gemini VI mission. "It is a treat for us to be here to receive these awards at the home of the president of the United States—the man who did so much to start this program out," he says. "I am sure he will be as interested as we are to see this lunar mission come to the final goal of sending three men to the Moon and bringing them back."

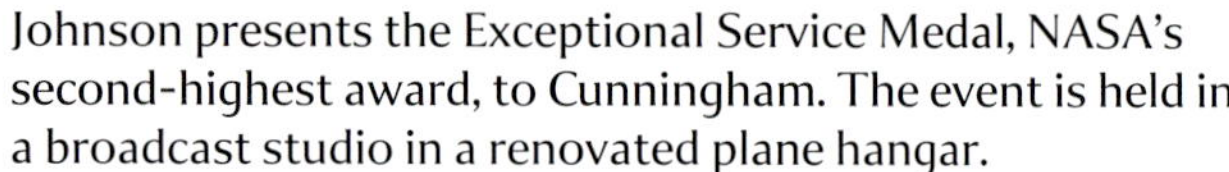

Johnson presents the Exceptional Service Medal, NASA's second-highest award, to Cunningham. The event is held in a broadcast studio in a renovated plane hangar.

Eisele receives his medal. Just-retired former NASA administrator Jim Webb receives NASA's highest honor, the Distinguished Service Medal.

The president watches a news conference with the crew, which is televised live. The astronauts are joined at the table by acting NASA administrator Tom Paine and Gilruth. NASA public-affairs chief Julian Scheer is at the lectern. *Left to right*: Lady Bird Johnson (*behind the president*); Johnson; Jo, Suzanne, and Marty Schirra; and Schirra's parents. US representative Olin Teague (D-Texas) is seated (*at center*). Schirra blames his "feisty attitude in orbit" on his cold, adding, "I do apologize to everyone for some of my remarks as they were phrased, but not for my decisions." The session is followed by a screening of a mission highlights film narrated by Schirra.

Schirra points out a detail in a photo of the S-IVB rendezvous over Texas to the president, as Cunningham removes his microphone.

Schirra and his crew thank the president before a coffee-and-cookies reception.

On November 3, the astronauts tape a segment for a *Bob Hope Special* before an invited audience of eight hundred in MSC's Building 2 auditorium. *Left to right*: MSC director of public affairs Paul Haney, Schirra, Cunningham, Eisele, Hope, and Barbara Eden, star of the TV series *I Dream of Jeannie*, which is set at Cape Kennedy. Schirra had received an offer to appear on the show. Eden gives them cards as honorary members of the TV and radio actors' union.

Left to right: Schirra, Cunningham, and Eisele. The special was timed to mark NASA's tenth anniversary and airs on NBC in prime time on November 6.

Schirra returns Hope's car key, after Cunningham tells him the astronauts would be taking Eden out that night, not Hope.

The crew and Hope then run through a skit featuring the comedian as their trainer, who gives them tips on how to make their TV transmissions more exciting. "Watch it," Hope tells them. "Nobody likes a wise astronaut." *Photo by Larry Evans* / Houston Chronicle

On November 6, Schirra and Cunningham land by helicopter at NAR in Downey, California.

"We have a very good spacecraft . . . a very good ride," Schirra tells the NAR workers. "Thanks to you people, it was no ordeal at all." Guests on stage include *Peanuts* cartoonist Charles Schulz (*second from left*) and Buzz Hello, Space Division VP (launch operations) (*third from left*).

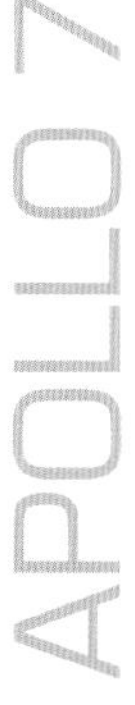

"We spent eleven very relaxed and pleasant days in space," Cunningham says. At right are William Bergen, Space Division president; John Moore, Aerospace and Systems Group president; John Atwood, president and CEO; and Dale Myers, Space Division VP.

Cunningham and Schirra head toward Building 247.

Schirra signs autographs for NAR workers. J. P. Michaels (*right*) wears his CM 101 badge.

Schirra autographs a Turtle poster during his visit.

An NAR executive speaks to Schirra and Cunningham as they inspect their CM.

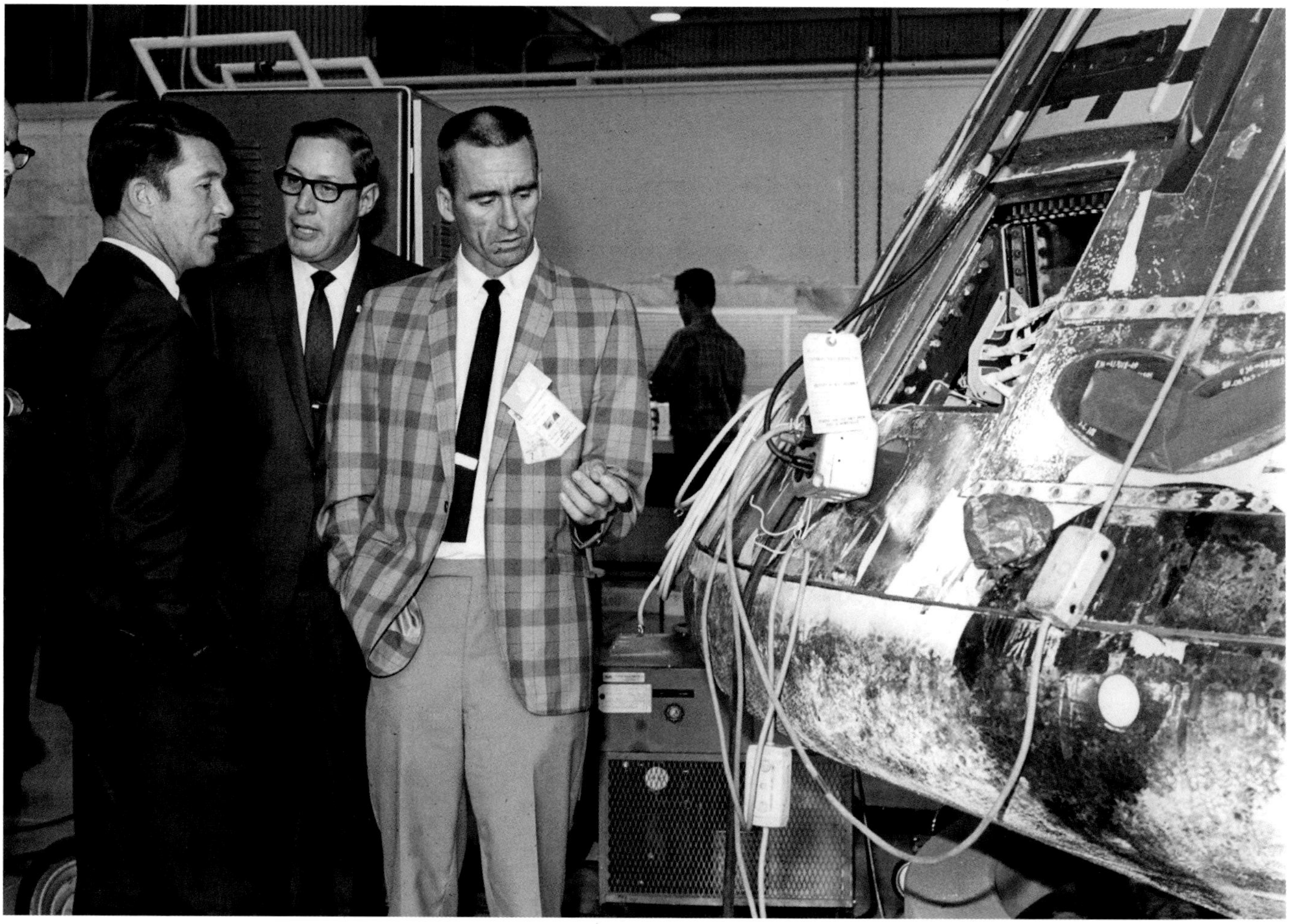

Schirra peers into the Apollo 10 CM, which will be sent to KSC later in the month.

Schirra and Cunningham at NAR

On December 9, aviation legend Charles Lindbergh, the first aviator to cross the Atlantic nonstop (*second from right*), signs a piece of stationery in the Treaty Room at the White House as the Apollo 7 and 8 astronauts watch. *Left to right, front row*: Cunningham, Eisele, Schirra, Bill Anders, Jim Lovell, Lindbergh, and Frank Borman; *back row*: First Lady Lady Bird Johnson, President Johnson, recently retired NASA administrator Jim Webb, and Vice President Hubert Humphrey. The stationery will join two other astronaut-signed pieces of paper hanging in the room, obtained by Jacqueline Kennedy (the Mercury 7) and Lady Bird (Gemini 4 crew).

Schirra smiles with the first couple behind him. The group next goes downstairs for a formal dinner honoring Webb, and Johnson awards him the Presidential Medal of Freedom, the nation's highest civilian honor. The 140 guests include seventeen other Apollo astronauts, Wernher von Braun, and famed aviator Jackie Cochran. *Bill Taub archive*

Eisele with a White House guest book. *Bill Taub archive*

Workers prepare a float with the Apollo 7 CM and an LM mockup for President Richard Nixon's inaugural parade on January 20, 1969. The three crewmen will ride in a convertible preceding the float. *Bill Taub archive*

On June 17, 1969, the astronauts are in Denver for a joint national meeting of the American Astronautical Society and the Operations Research Society of America. Cunningham tells a news conference, "We need a breakthrough in propulsion before we can really go about the job of investigating our solar system." Schirra was the keynote speaker.
Photo by Duane Howell / Denver Post

Epilogue

Schirra smiles on December 17, 1968, after his wife, Jo, pins the Navy Distinguished Service Medal on his lapel. He receives the award for commanding Apollo 7 during a ceremony at the Pentagon in Arlington, Virginia. "I think we'll demonstrate that [Apollo 8] is ready. It's just a little higher and a little farther," he says of the Moon-orbiting mission set for launch in five days.

After joining CBS News as an analyst in July 1969, Schirra poses with models of the Atlas, Titan, and Saturn IB boosters he's flown, and a Saturn V model. He had resigned from NASA on July 1, and all three TV networks had been interested in signing him for their on-air coverage. *CBS photo*

Schirra (*right*) with anchor Walter Cronkite on the CBS News Space Center set in New York after the Apollo 11 Moon landing on July 20, 1969. Schirra would continue in the role into 1975.

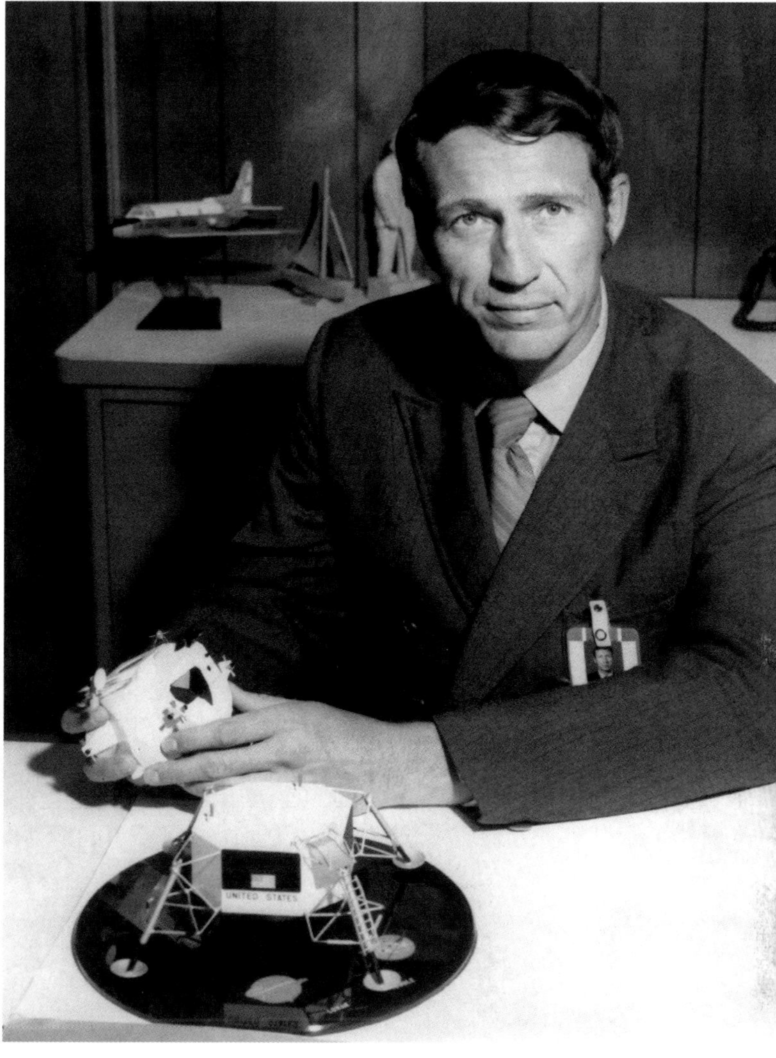

Eisele's last NASA portrait is released in 1971. He had resigned from the Astronaut Office in 1970 to become technical assistant for manned spaceflight at the Langley Research Center in Hampton, Virginia. Among future projects he works on is a reusable space shuttle.

Eisele with a Grumman LM model at Langley in 1971

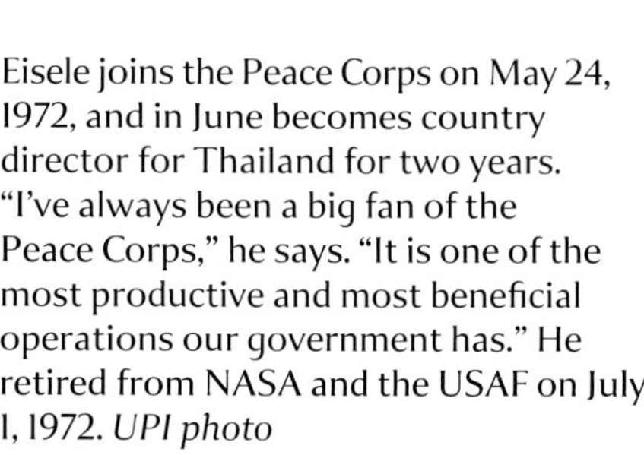

Eisele joins the Peace Corps on May 24, 1972, and in June becomes country director for Thailand for two years. "I've always been a big fan of the Peace Corps," he says. "It is one of the most productive and most beneficial operations our government has." He retired from NASA and the USAF on July 1, 1972. *UPI photo*

Cunningham is greeted by deputy director William Lucas (*right*) during a visit to Marshall Space Flight Center on January 28, 1970, with MSFC director Eberhard Rees (*at left*). Cunningham heads the Skylab branch of the Flight Crew Directorate at MSC. Marshall is responsible for the design, manufacture, and assembly of Skylab's Apollo Telescope Mount.

Astronauts attend a crew station review of the Skylab Airlock Module and Multiple Docking Adapter at McDonnell Douglas in St. Louis on July 29, 1970. *Left to right*: Al Bean, Pete Conrad, Don Holmquest, Don Lind, Dick Truly, Cunningham, Rusty Schweickart, and Bob Overmyer. Cunningham is initially promised command of the first manned mission to Skylab but is bumped by Conrad's seniority that fall.

Cunningham announces his resignation from NASA on June 17, 1971, at MSC. "The shortsighted, low-level support of the space program and the future best interests of my family have convinced me this is the right move for us at this time," he tells reporters. He also cites losing the Skylab mission.

The starboard side of the Apollo 7 CM at the KSC Visitor Center in January 1974. After appearing in President Nixon's inaugural parade in January 1969, the spacecraft was displayed for five years at KSC, on loan from the Smithsonian Institution. It was then loaned to the National Museum of Science and Technology in Ottawa, Ontario, later that year. *Photo by J. L. Pickering*

The Apollo 7 CM hatch at KSC. *Photo by J. L. Pickering*

The port side of the CM at KSC in January 1974. *Photo by J. L. Pickering*

In April 2004, the Apollo 7 CM returned to the US for display at the Frontiers of Flight Museum in Dallas, Texas, seen in October 2017. *Photo by Mark Usciak*

The unified hatch is open for the display at the museum, seen in October 2017. *Photo by Mark Usciak*

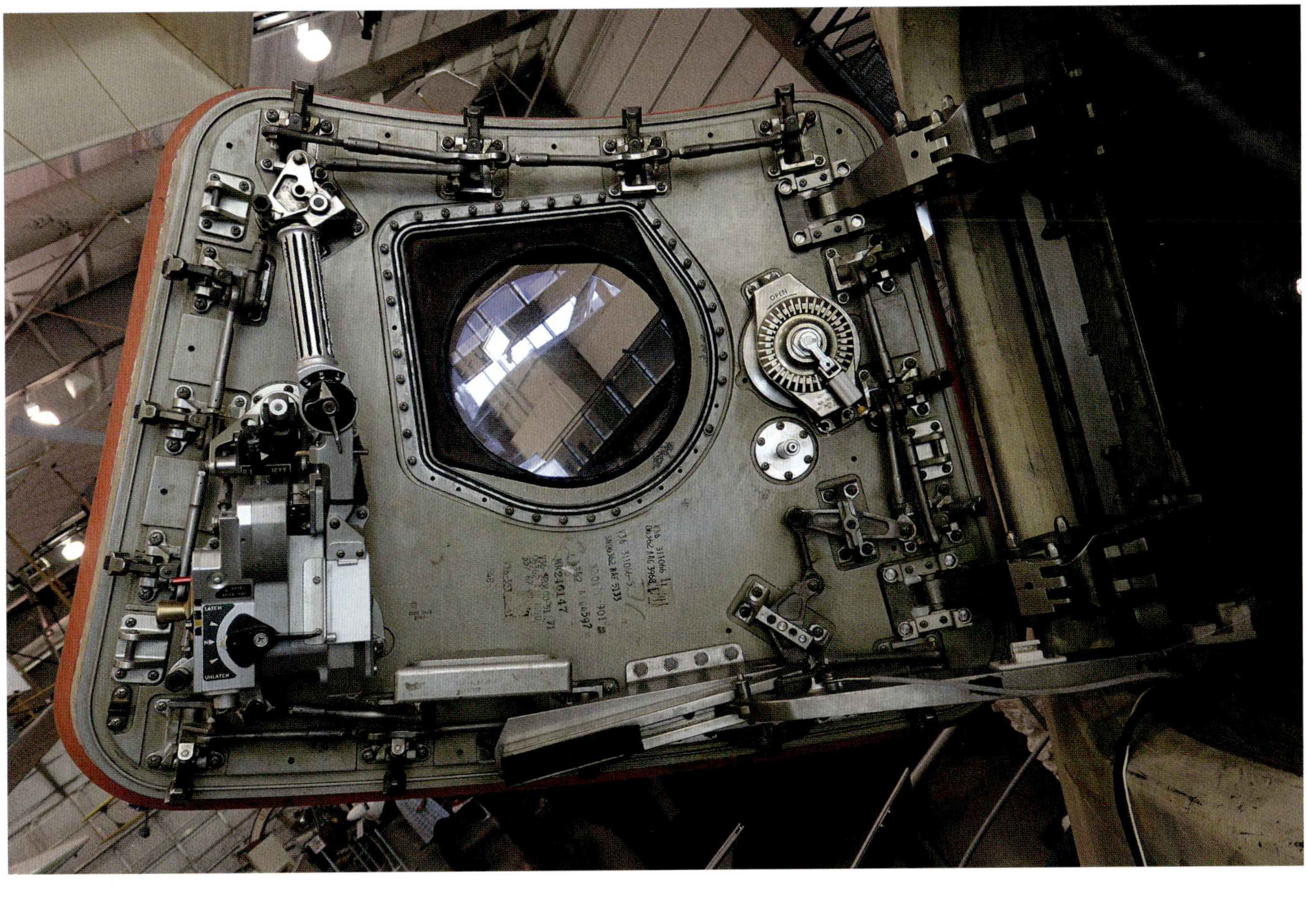

The honeycomb structure of the aft heat shield is visible, with small brown plugs filling locations where core samples were taken. Photo taken in October 2017.
Photo by Mark Usciak

Visitors can view the lighted crew compartment of the CM in this photo from July 2018 at the Frontiers of Flight Museum.

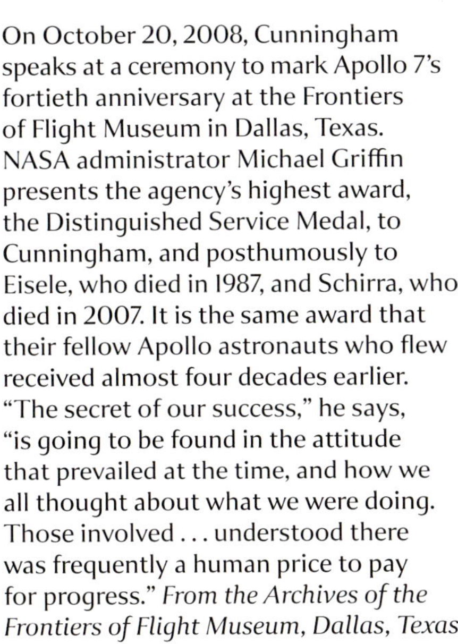

On October 20, 2008, Cunningham speaks at a ceremony to mark Apollo 7's fortieth anniversary at the Frontiers of Flight Museum in Dallas, Texas. NASA administrator Michael Griffin presents the agency's highest award, the Distinguished Service Medal, to Cunningham, and posthumously to Eisele, who died in 1987, and Schirra, who died in 2007. It is the same award that their fellow Apollo astronauts who flew received almost four decades earlier. "The secret of our success," he says, "is going to be found in the attitude that prevailed at the time, and how we all thought about what we were doing. Those involved . . . understood there was frequently a human price to pay for progress." *From the Archives of the Frontiers of Flight Museum, Dallas, Texas*

Cunningham visits with astronauts Bill Anders and Buzz Aldrin. Also in attendance are astronauts Neil Armstrong, Alan Bean, and flight director Gene Kranz. *From the Archives of the Frontiers of Flight Museum, Dallas, Texas*

Cunningham and his wife, Dot, pose with one of more than thirty donated Saturn IB models to be auctioned to benefit Be An Angel, a charity they are involved with that helps children with disabilities. Cunningham died in 2023. *From the Archives of the Frontiers of Flight Museum, Dallas*

Abbreviations

AFB	air force base
AFS	air force station
AMS	Apollo Mission Simulator
AS	Apollo-Saturn
BP	boilerplate
CDDT	Countdown Demonstration Test
CM	command module
CMPS	Command Module Procedures Simulator
CSM	command/service module
EDT	Eastern Daylight Time
EST	Eastern Standard Time
EVA	extravehicular activity
IU	instrument unit
KSC	Kennedy Space Center
LC	launch complex
LES	launch escape system
LM	lunar module
LOX	liquid oxygen
MA	Mercury-Atlas
MOCR	Mission Operations Control Room
MSC	Manned Spacecraft Center
MSFC	Marshall Space Flight Center
MSOB	Manned Spacecraft Operations Building
MV	motor vessel
NAA	North American Aviation
NAR	North American Rockwell
NASA	National Aeronautics and Space Administration
PAA	Pan American World Airways
RCS	reaction control system
SA	Saturn
SLA	Spacecraft / Launch Vehicle Adapter
SM	service module
SPS	service propulsion system
US	United States
USAF	United States Air Force
USN	United States Navy
VAB	Vehicle Assembly Building

Bibliography

Apollo 7 Mission Evaluation Team. *Apollo 7 Mission Report*. MSC-PA-R-68-15. Houston, TX: Manned Spacecraft Center, December 1968.

"Apollo 7's Big Splash in Color." *Broadcasting*, October 28, 1968, 72. Washington, DC: Broadcasting Publications.

Benson, Charles D. *Moonport: A History of Apollo Launch Facilities and Operations*. Washington, DC: Scientific and Technical Information Office, National Aeronautics and Space Administration, 1978.

Brooks, Courtney G., and Ivan D. Ertel. *The Apollo Spacecraft: A Chronology*. Vol. III. Washington, DC: National Aeronautics and Space Administration, 1976.

Chrysler Corp. Space Division. *Saturn IB Orientation Systems Training Manual 851-0*. 1965.

Coan, Paul P. "Apollo Experience Report: Television System." NASA Technical Note D-7476. Houston, TX: National Aeronautics and Space Administration, Lyndon B. Johnson Space Center, 1973.

Department of the Interior, National Park Service, Southeast Region. *Cape Canaveral Air Force Station, Launch Complex 39, Altitude Chambers*. Atlanta: Historic American Engineering Record, 2009.

Ertel, Ivan D., and Mary Louise Morse. *The Apollo Spacecraft: A Chronology*. Vol. I. Washington, DC: National Aeronautics and Space Administration, 1969.

General Motors Corp., AC Electronics Division. *Apollo Guidance and Navigation Student Study Guide—Block I (Series 100) G&N System Familiarization*. 1965. Revision B, 1966.

Gibson, Walter G. "A Scan Converter for Apollo Television." *RCA Engineer* 11, no. 6 (April–May 1966). Princeton, NJ: Radio Corporation of America.

Kozloski, Lillian D. *US Space Gear: Outfitting the Astronaut*. Washington, DC: Smithsonian Institution Press, 1994.

Morse, Mary Louise, and Jean Kernahan Bays. *The Apollo Spacecraft: A Chronology*. Vol. II. Washington, DC: National Aeronautics and Space Administration, 1973.

National Aeronautics and Space Administration. *Apollo 7 Mission Commentary*. Houston, TX: Manned Spacecraft Center,1968.

National Aeronautics and Space Administration. "Apollo 7 Press Kit." Release No. 68-168K. Washington, DC: National Aeronautics and Space Administration, October 1968.

National Aeronautics and Space Administration (Marshall Space Flight Center, Kennedy Space Center), Chrysler Corp. Space Division, McDonnell Douglas Astronautics Company, IBM Federal Systems Division, and Rocketdyne. *Saturn IB News Reference*. 1965.

National Aeronautics and Space Administration, Launch Operations Center. "NASA Space Utilization at AMR." Cape Canaveral, FL: National Aeronautics and Space Administration, Launch Operations Center, 1962.

National Aeronautics and Space Administration. *Apollo Recovery Operational Procedures Manual, Revision A*. Houston, TX: Manned Spacecraft Center, Landing and Recovery Division, 1969.

National Aeronautics and Space Administration. "Launch Complex 34 Facilities." KSC Fact Sheet 05. 1968.

North American Aviation, Space and Information Systems Division. *Apollo Operations Handbook—Command and Service Module—Spacecraft 012*. Downey, CA: North American Aviation, Space and Information Systems Division, 1966.

North American Rockwell. *Apollo Spacecraft Familiarization Manual*. Downey, CA: North American Rockwell, 1967.

North American Rockwell, Space Division. *Apollo Spacecraft News Reference*. Downey, CA: North American Rockwell, 1969.

Pavlosky, James E., and Leslie G. St. Leger. *Apollo Experience Report Thermal Protection Subsystem*. Houston, TX: National Aeronautics and Space Administration, Lyndon B. Johnson Space Center, 1974.

Phinney, William C. "Science Training History of the Apollo Astronauts." NASA SP-2015-626.

Saturn V Launch Vehicle Flight Evaluation Report—AS-502 Apollo 6 Mission. Prepared by Saturn V Flight Evaluation Working Group, George C. Marshall Space Flight Center June 25, 1968, NASA MPR-SAT-FE-68-3.

Slovinac, Patricia. *Cape Canaveral Air Force Station, Launch Complex 39, HAER No. FL-8-11-E Altitude Chambers*. Atlanta: Historic American Engineering Record, National Park Service Southeast Region, Department of the Interior, 2009.

Index

Note: the following entries are too numerous to include: Apollo 7; Cape Kennedy AFS, command module; Cunningham, Walt; Eisele, Donn; LC–34; NASA; Schirra, Wally; Saturn IB; S-IB, S-II, S-IVB, service module.

VII
SCHIRRA · EISELE · CUNNINGHAM